NCC
The National Computing Centre

The National Computing Centre develops techniques and provides aids for the more effective use of computers. NCC is a non-profit-distributing organisation backed by government and industry. The Centre

- co-operates with, and co-ordinates the work of, members and other organisations concerned with computers and their use
- provides information, advice and training
- supplies software packages
- publishes books
- promotes standards and codes of practice

Any interested company, organisation or individual can benefit from the work of the Centre by subscribing as a member. Throughout the country, facilities are provided for members to participate in working parties, study groups and discussions, and to influence NCC policy. A regular journal – 'NCC Interface' – keeps members informed of new developments and NCC activites. Special facilities are offered for courses, training material, publications and software packages.

For further details get in touch with the Centre at Oxford Road, Manchester M1 7ED (telephone 061-228 6333)

or at one of the following regional offices

Belfast
1st Floor
117 Lisburn Road
BT9 7BP
Telephone: 0232 665997

Birmingham
2nd Floor
Prudential Buildings
Colmore Row
B3 2PL
Telephone: 021-236 6283

Bristol
6th Floor
Royal Exchange Building
41 Corn Street
BS1 1HG
Telephone: 0272 27077

Glasgow
2nd Floor,
Anderston House
Argyle Street
G2 8LR
Telephone: 041-204 1101

London
11 New Fetter Lane
EC4A 1PU
Telephone: 01-353 4875

Selection of Data Communications Equipment

PUBLISHED BY NCC PUBLICATIONS

Keywords for information retrieval (drawn from *NCC Thesaurus of Computing Terms):* data transmission equipment; data transmission systems; data transmission.

British Library Cataloguing in Publication Data

Nichols, E
Selection of data communications equipment.
1. Data transmission systems
I. Title II. Jocelyn, S
621.38'028 TK5105

ISBN 0-85012-217-1

Published in July 1979 by:
NCC Publications, The National Computing Centre Limited,
Oxford Road, Manchester M1 7ED,

Printed in England

ISBN 0-85012-217-1

Selection of Data Communications Equipment

Prepared for The National Computing Centre Ltd by:

Management Group
LOGICA LTD
PO Box 4SE
64 Newman Street
London W1A 4SE

Authors: Eirwen Nichols, Sarah Jocelyn

Edited by: Lyndon Morgan (NCC)

Preface

Any organisation moving into the field of data communications has the problem of selecting the correct hardware. The hardware must meet current requirements and be compatible with future needs. The data communications hardware market is very large. Suppliers range from small companies to large multinationals. Products are numerous with new items appearing regularly.

There is no shortage of information on data communications hardware: product reviews are to be found in many periodicals. However, it was felt that what was lacking was a straightforward guide to a rigorous method of data communications hardware selection. Such a guide would allow a computer manager to make a quantitative assessment of data communications needs and evaluate the currently available equipment against those needs.

It is hoped that this book will facilitate the selection of data communications hardware, with decisions made on a quantitative basis with due consideration of all the factors involved.

Contents

1 Introduction

OBJECTIVES

The increasing acceptance of the computer in industry and government has been coupled with an explosive growth in the use of data communications. The indications are that this growth will continue well into the 1980s: soon the computer without any data communications facilities will be the exception. However, the speed with which data communications technology has developed, the proliferation of hardware and software, and the associated jargon, have created the impression that the subject is one best left to the specialist.

In the case of a large sophisticated network this is undoubtedly true. Here the investment in design and installation is considerable and the wrong choice of transmission speed or hardware, for instance, could have serious and expensive repercussions. More and more, however, simple networks are being used by organisations to connect a few remote terminals to a central computer.

The installation or expansion of such a system can often provide a good opportunity for data processing staff to become more familiar with data communications. Inevitably, much of the hardware will be located in the computer department, and it is usually most convenient for a member of this department to undertake responsibility for network management. The sooner practical experience is gained the better.

This book is intended primarily for readers operating in either of two particular environments (although it should also be of general use to readers already working with a complex network):

– a totally batch environment which is to be modified to include a simple data communications system;

– an environment where a simple data communications system is already operational and is to be extended by the addition of a small number of terminals.

In both cases the expenditure is not likely to be large enough to justify the use of consultants unless the systems contain unusual design features.

The selection of data communications hardware – the area of heaviest capital investment – is one of the most critical tasks involved in the design and installation of such a system. The large range of hardware, of varying levels of sophistication and at widely varying costs, can make this task appear daunting to the inexperienced data processing or operations manager. This book aims to provide information and guidance on the selection of equipment so that the average processing or operations manager with no previous data communications experience can

– select the necessary hardware for a fairly simple installation;

– understand better the hardware requirements and the facilities available to allow more effective use of any specialist communications staff employed to design and implement a system.

Such knowledge will also be of use, once the network is operational, as a basis for efficient network control, monitoring and fault-finding.

SCOPE

The book is aimed at senior data processing personnel and therefore assumes a detailed understanding of data processing terms and concepts. On the other hand, little previous knowledge of data communications is assumed and a chapter is devoted to the relevant principles of data communications. Emphasis is placed on the hardware aspect and a reader interested in the more general principles is recommended to consult an appropriate publication, such as the *Handbook of Data Communications* (NCC Publications).

The following types of hardware are discussed:

– modems;

– multiplexers;

– concentrators;

- communications control units and front-end processors;
- test and diagnostic equipment;
- racks;
- switching and patching equipment.

This list includes all the major items of data communications equipment available today.

The book aims to define each of these items, to explain when and how they are required, and to draw up a list of the criteria by which each can be judged. First, the necessary steps in network design are discussed so that all the factors which affect selection are covered.

A straightforward weighted ranking method is proposed to facilitate objective selection of each item. The hardware is discussed with the evaluation method in mind. It is up to the reader to judge the relative importance of any criterion in selection, as this will depend on individual circumstances. The application of the evaluation method to certain equipment, such as a concentrator (where the manufacturer will to a large extent tailor the device to a customer's requirements), is less meaningful than in the selection of off-the-shelf devices.

Overall considerations which must be borne in mind once the weighted ranking evaluation has been carried out are discussed in Chapter 10.

A case study is used to demonstrate the practical application of the evaluation method to the selection of hardware for a network which is planned to develop over a five-year period. Practical decisions in equipment selection are also highlighted, such as the selection of equipment which will be of use throughout the various stages of network development.

The user needs to be able to select hardware and to know which manufacturers can supply it. To assist in this, an appendix lists sources of information on equipment suppliers.

2 Principles of Data Communications

INTRODUCTION

Data communications is based on data processing and on data transmission techniques. There is a considerable amount of basic information and jargon with which the reader must be familiar before a simple network can be designed. The aim of this chapter is to provide the essential definitions and concepts needed for a specification of hardware requirements. All aspects of a data communications network are covered, but in keeping with the theme of the book, only those related to hardware selection are discussed in depth. Readers interested in studying the whole subject further should consult appropriate books (see Bibliography).

DEVELOPMENT OF DATA COMMUNICATIONS

Data communications is defined in the *Handbook of Data Communications* (NCC Publications) as embracing not only the electrical transmission of information in coded form but other factors in controlling, checking and handling the movement of information in a communications-based computer system. The growing trend for the distribution of intelligence throughout a network makes the boundary between data processing and data communications extremely blurred. The US Federal Communications Commission, responsible for monitoring and proposing legislation in the field of communications, has discussed the subject for some time with no firm conclusions reached.

Data communications, based on the partnership between computers and electrical transmission, began in the mid-sixties. Computers became increasingly sophisticated and hardware prices began to fall, so people began to think of extending the power of the computer beyond the computer room and to allow terminals at remote locations to access it. This trend has grown steadily and it is predicted that by 1990 nearly all computers will have some data communications facilities.

Electrical transmission of the human voice has been with us for over a century and the facilities available for this, ie the Public Switched Telephone Network (PSTN), are well established. It was therefore natural for this and the other well established service for message transmission, the telegraph, to be considered for data transmission applications. However, telephone facilities were designed for transmission of analogue waveforms, a means of converting digital data to analogue format for transmission, and back to digital format for processing, was required. Use of the Post Office telephone plant in this way has proved successful though not ideal, especially at medium and high speeds. Use of the telegraph service has been limited by its low speed. As the volume of data transmission has grown, increasing attention has been given to the provision of special facilities for data.

In all European countries, the provision of facilities for data communications is the responsibility of the national Postal and Telegraph authorities, generally referred to as the PTTs. In Britain the responsible body is the Post Office.

Public and Private Data Network

Recent years have seen the growth of both *public* and *private* data networks. A private network is one created exclusively for the use of one organisation, a public network being one established by the PTT (the best known example being the public telephone network).

In addition to making use of the public telephone and telex networks, it is possible for an organisation to lease lines (for use in data transmission) from the national PTT. In this way organisations, both large and small, have built up their own private data networks which operate both nationally and internationally. These networks vary in size and sophistication, and range in use from a real-time airline reservation system to a simple batch-type stock control system. Many organisations in fact create networks of both private lines and PSTN facilities, the choice depending on economics and convenience.

As data communications traffic has grown over the past few years the PTTs have responded to demand by developing public networks exclusively for data transmission. Some European countries are more advanced than others but by 1982 most countries in Europe will have such a network although not all will be based on the same technology. France, for example, will have two public data networks. One (known as Caducee) became operational first, and the other (known as Transpac) became operational in 1979.

A further development, taking place in North America and soon to be noticed in Europe, is the provision of digital links for data. These will give users far better error rates, especially at higher transmission speeds.

On-Line and Off-Line Systems

In data communications systems the telecommunications lines may be *on-line* or *off-line* to the computer. On-line data systems are those where the lines are connected to the host computer, which itself controls transmission. The data may not be processed immediately but the data is stored ready in the host for processing as necessary. An off-line system is one which is isolated from the host computer. The data is transmitted to the host site and written to a magnetic tape, punch cards or paper tape. This is then physically transferred to the host computer which must read it before any processing is possible. Both types of system are shown in Figure 2.1.

Off-line data communications, relatively cheap and simple, was used extensively in the first data communications systems. It is especially applicable where there is a need to move large quantities of data from one location to another. In general however, on-line systems offer the user greater convenience and speed. They are usually employed where high response times are required, or where there is a requirement for validity checks to be made in transactions at the time they are entered. The hardware required for both on-line and off-line systems is the same with one exception – no front-end processor or communications control unit is necessary in an off-line system.

Distributed Processing

The early data communications networks of the 1960s were simple in structure. Usually there was a remote terminal or terminals connected to a mainframe computer which carried out all the information processing. By the 1970s, with the development of both computer and telecommunications technologies, it

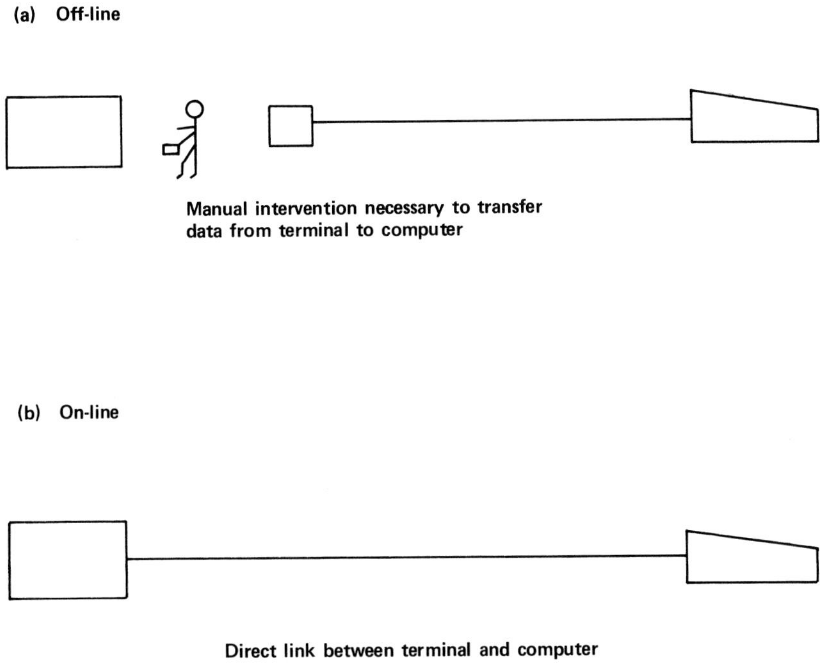

Figure 2.1 On and Off-line data transmission

has become possible for the information processing to become dispersed throughout the network.

'Intelligent' or programmable terminals, as well as equipment based on minicomputers (eg concentrators), have been developed, all able to carry out some information processing previously performed by the mainframe computer. This 'distributed processing' offers certain distinct advantages, such as the location of processing power where it is needed and increased security from decentralised operation, but at the expense of increased complexity and initial investment. (The interested reader is referred to two NCC publications: *Trends in Distributed Processing* and *Why Distributed Computing?)*

Applications of Data Communications

The variety of data communications network applications is enormous. Networks have been set up for scientific, educational, military, commercial, government and leisure purposes. Common applications include stock control, order-entry, reservation systems, payroll, financial and management reporting, and interactive computational applications.

The speed with which information can be collected and analysed by a computer via a data communications system is its prime advantage. It allows the organisation to make decisions based on more up-to-date information and gives better control. In addition the ease of use of a well designed system can improve the efficiency of an organisation. For example, a clerk may feed data directly into the computer via a terminal at a remote location, instead of filling it onto a form which is dispatched to the computer centre and then fed into the computer by another person.

HARDWARE

Elements of a Data Communications System

Any data communications network comprises three elements:

– *Host computer,* which provides services to a remote user via network facilities and carries out central processing functions;

– *Terminal equipment,* ie any input/output device allowing a user access to host computer facilities. It may or may not have the ability to carry out some processing itself. Such equipment is often referred to as Data

Terminal Equipment (DTE);

- *Communications links and interfaces,* ie equipment that allows terminal equipment and hosts to communicate.

This book concentrates on the third element. Frequent reference will be made to host computers and terminal equipment, but the main aim is only to highlight the factors which are important to the selection of communications hardware.

The hardware required in a network depends on how each terminal is to be connected to the host (or hosts) to which it requires access. There are several solutions to this problem. In increasing order of complexity these are:

- point-to-point connections;
- multipoint or multidrop connections;
- use of a multiplexer;
- use of a concentrator.

These are discussed below; first, attention is given to communications links themselves and the means of data transmission.

Features of a Communications Link

Public, Telegraph or Private Link

A user in Britain has three alternative paths for data transmission: a public telephone circuit, a telegraph circuit, or a private telephone link. All three must be supplied by the Post Office.

Excluding the low-speed telegraph facilities, there are neither digital data links nor a public data network in Britain at present, though the latter is planned to be operational by 1980. Use of a public or PSTN circuit is frequently referred to as 'dial up'. This is because when a user wishes to transmit data between terminal and computer he establishes the connection or 'sets up the call' in exactly the same way as he would make a normal phone call, ie by dialling the computer's 'telephone number'. The connection remains until the caller breaks it by hanging up.

As with a normal telephone call, the user is charged on a per minute basis. In addition there is an installation charge and a monthly rental. Error rates over the PSTN tend to be higher than over a private circuit, especially at higher speeds, primarily because of the number of switching exchanges through which the call is routed. However, as the quality of data transmission equipment is improved, the speed at which data may be transmitted over the PSTN is increased.

The public telegraph or telex network, a digital network, is also available for transmission of data in a similar way to the PSTN but only at low speed, viz 50 bps. It is therefore not used extensively.

A private line, alternatively called a leased or dedicated line, may be leased from the Post Office for the exclusive use of an organisation. This is a permanent connection and, once a terminal is switched on, it can transmit to the equipment at the other end immediately. There is a fixed monthly charge but no usage charge so such a line becomes economically advantageous between sites which exchange large volumes of data.

The Post Office offers a range of facilities, not just lines, for data transmission. These are discussed later.

Two- or Four-Wire Circuits

A line may be two-wire or four-wire. The former provides one channel which is generally only usable in one direction at a time; the latter provides two independent channels, permitting transmission in both directions simultaneously. Four-wire circuits are usually used by the Post Office for distances over approximately 40 kilometres.

Transmission Features

Duplex, Half-duplex and Simplex Transmission

There are four ways of transmitting data:

– Simplex: A two-wire circuit is used to transmit data in one direction only, ie A to B is possible but B to A is not. This method is not used frequently in data communications. One application is in telemetry where a remote station sends information to a computer centre for monitoring and control purposes;

– Half-duplex: A two-wire circuit is used to send information in both directions but not simultaneously, ie A to B, and also B to A, are possible but not at the same time. The line must be 'turned around' to permit the transmission of data from B to A once transmission from A to B is complete. This is carried out automatically by the modems at each end of the line.

– Duplex or Full-duplex	A four-wire circuit is used to transmit data in both directions simultaneously, ie, A to B, and B to A, are possible at the same time. One pair of wires acts as the transmit channel and the other pair as the receive channel;
– Asymmetric Duplex (or 2 Full-duplex)	A two-wire circuit can be used to transmit data in both directions simultaneously by dividing up the channel. This may be done so that two low-speed channels (eg 300 bps) are provided or one medium-speed channel (eg 600/1200 bps) and one low-speed channel (eg 75 bps). In the latter case, the low-speed channel is sometimes called the 'reverse' or 'backward' channel. This type of transmission is not widely used. It is useful for information control purposes. A recent application is however the Post Office's Prestel service (Viewdata information system). The viewer requests the information to be displayed on his television screen using the low-speed channel. The high-speed channel is then used to relay the mass of information to be displayed on the screen.

The terms simplex, half-duplex, duplex and asymmetric duplex are sometimes used differently to the definitions given above, although these definitions given here are the most common. It is therefore worthwhile checking exactly what is meant by these terms when they are encountered.

Synchronous and Asynchronous Transmission

Data is transmitted along the line as a sequence of characters, each character comprising a number of bytes or words (group of '1' and '0' bits).

Some method is required to indicate the beginning and end of each character and to show which bit is which in the character sequence, so that the data transmitted along the line will be received at the other end in the same format that it was sent.

There are two ways of transmitting data between terminal and computers.

– Asynchronous Transmission

In this kind of transmission data is sent to the receiver one character at a time. The characters are transmitted in a start and stop fashion. A start bit defines the beginning of the character and a stop bit defines the end of a character. These start/stop bits do not themselves convey any data, they simply indicate that a byte is about to be transferred. Thus synchronisation of the transmitting and receiving devices occurs on a character-by-character basis. Once the stop bit of the character has been received, the receiving device, whether it be the terminal or host computer, will stop and wait for the start of the next character.

To effect this 'start-stop' method of transmission the terminals or computer have clocks or internal timing devices which are set so that the clock at the transmitting end is in time with the clock at the receiving end. The clocks at both ends are thus each set to define the successive short intervals of time which represent the bits transmitted.

Although asynchronous transmission is simple to operate it is an inefficient transmission method. For every character sent, extra start-stop characters are required for synchronisation purposes. As the transmission speed increases, the total amount of data that could have been transmitted instead of stop-start bits becomes significant. For example at 300 bps, assuming stop-start bits occupy 20%, only 60 bits of significant data are lost per second, while at 4800 bps this becomes 900 bits.

It is, therefore, more suitable for sending small amounts of data at lower speeds. It is usually used at the lower transmission speeds, up to 1200 bps, on keyboard devices, such as teleprinters, which do not have a buffer and on which the operator keys in information at random intervals.

– Synchronous Transmission

With this method of transmission, data is transmitted as a continuous stream or block of characters instead of separate isolated characters. The start and stop bits for each character are thus eliminated. However, the transmitting and receiving terminal/computer clocks must still be in time with each other so that the beginning of each stream of characters can be identified by both ends. The start of each stream is shown by preceding each stream of characters with two or more bytes having a predetermined bit pattern. It is this predetermined synchronisation pattern which is recognised by the receiver as being the start of the character stream.

This method of transmission is more efficient for large amounts of data although it is a more costly process than asynchronous transmission as the circuitry within the modems as well as in the terminals is more complex. It is nearly always used at transmission speeds of 2400 bps and above.

Bauds and Bits Per Second (bps)

Rates of data transmission can be defined in two ways (frequently confused with each other and misused):

– Baud

  Bauds = a unit of modulation speed. This is the number of times per second that an information signal changes and is also called the modulation rate.

– Bits per second (bps)

  Bits per second is the figure which represents the true data transfer rate. It is the number of bits of information transmitted each second.

When every signal element transmitted represents one bit ('1' or '0') then the baud rate is the same as the number of bits per second. When a signal element represents more than one bit of information then the bit rate is higher than the baud rate.

Generally in data transmission techniques used below 2400 bps, the baud and bit rate are equal. To achieve speeds in excess of 2400 bps, techniques must be used where every signal element represents more than one bit. In this case for example the baud rate could be 2400 baud while the rate of data transfer could be 9600 bps, 4 bits of data being contained in each signal element.

Point-to-Point Connections

This is the simplest way to connect a terminal to a host computer. They are connected by a direct link (fig. 2.2). Each terminal connected in this way requires its own input/output port at the host, so this solution is only feasible for a small number of terminals. The link may be either a permanently connected leased line or a PSTN or telex link. Whichever method is used, once the connection is established it is used only for exchange of information between the terminal-host pair, similar to the way a telephone circuit is used exclusively for one conversation once a call has been established between two parties.

The terminal and computer in Figure 2.2(a) both require data in digital format, while the link is designed for analogue signals. Some device is therefore necessary to convert data from digital to analogue format and vice versa. Such a device is known as a MODEM (acronym for MOdulator/DEModulator) although the term DATA-SET, used widely in North America, may be used by US companies marketing their products in the UK. A modem is the most important piece of hardware in any data transmission network. One is required at the end of each link, as shown in Figure 2.2(b) which shows a physical representation of the logical connection shown in Figure 2.2(a). Modems are discussed in detail in Chapter 5.

Multipoint or Multidrop Links

The terms multipoint and mutidrop are frequently used synonymously but this is not strictly correct. Both allow more than one terminal to share a single line, the differences being shown in Figure 2.3. Signals can pass from the computer to any terminal on the line and vice versa but the individual terminals on the line cannot communicate with each other. If the terminals are not heavily used this can offer significant savings in line costs, although a leased line is necessary, such a feature not being possible via the PSTN. A modem is required at each line termination, ie at each terminal and at the host, as shown. A multidrop configuration is not possible in Britain, the Post Office provides only a multipoint line with two branching points.

On a multipoint line only one device may be transmitting at any

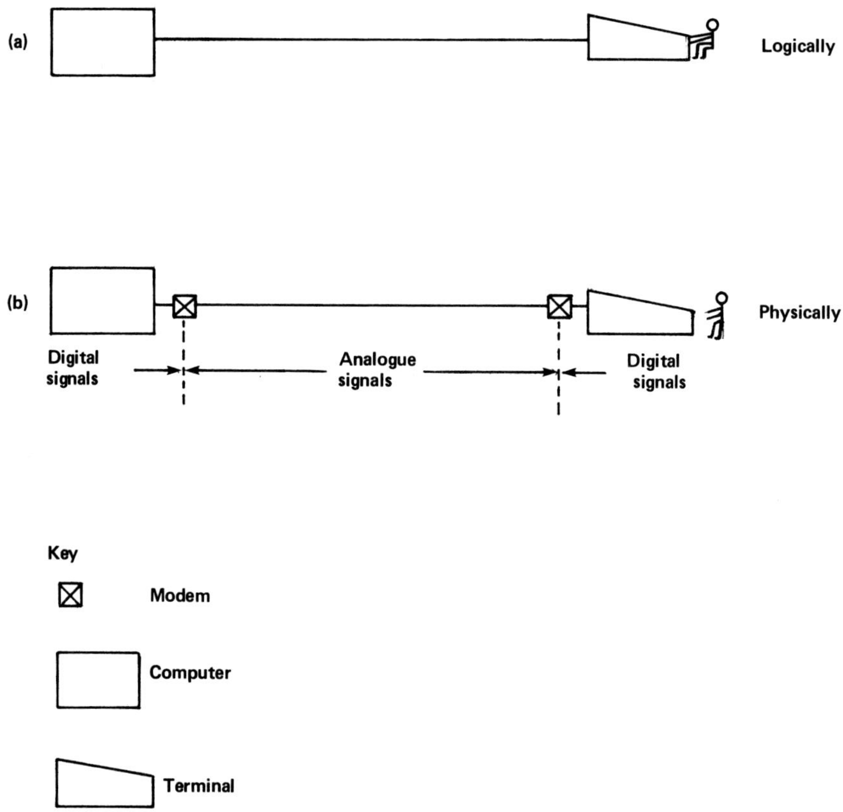

Figure 2.2 Point-to-point link

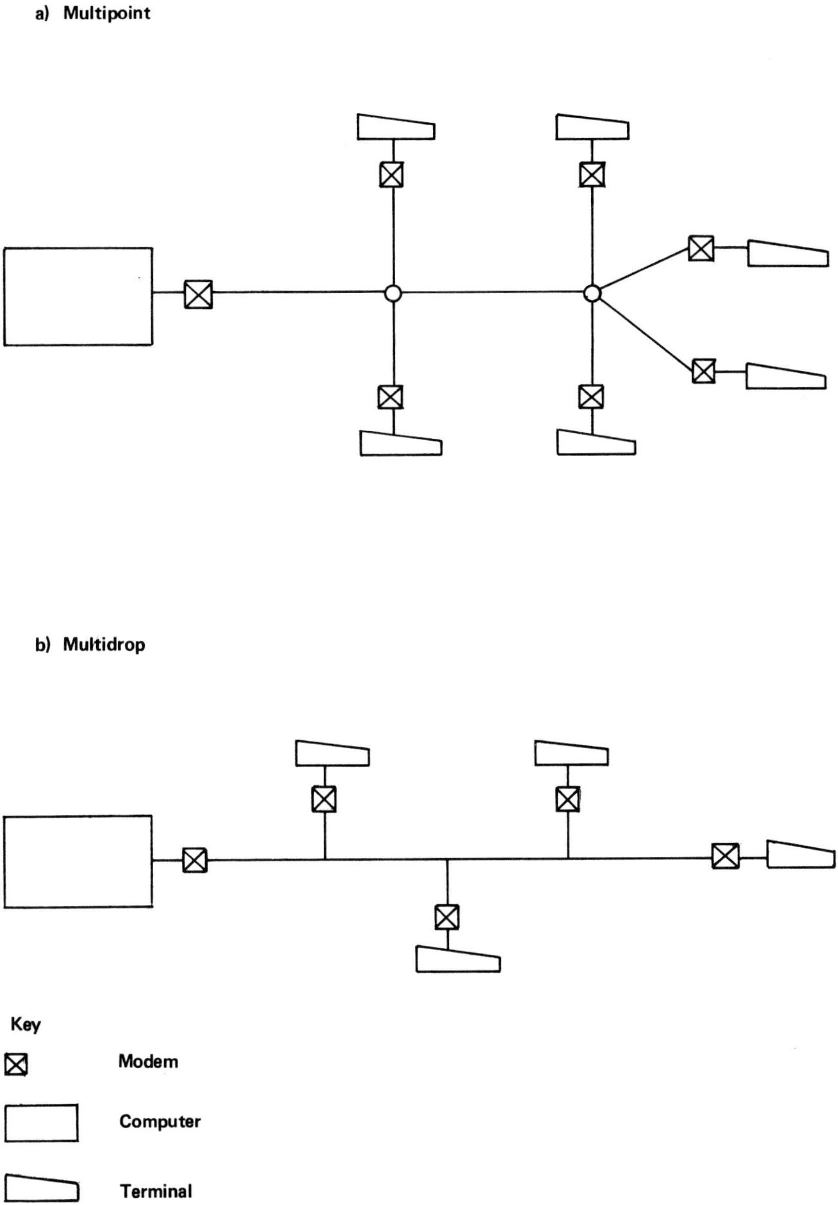

Figure 2.3 Multipoint and multidrop links

one time; the devices are thus contending for the line. This is organised by the host computer 'polling' or calling up each device in a predefined order and asking them to transmit data. This contention can cause a few seconds delay in that each terminal has to wait its turn and join the queues for sending and receiving data. The greater the number of devices on the line the longer this delay can be.

Multiplexers

Because of the increase in response time on a multipoint line as the number of terminals increases, other methods of allowing terminals to share a line have been developed. The most widely used of these is based on a device called a MULTIPLEXER. Many terminals can be connected to a multiplexer by pont-to-point links, the multiplexer then transmits the data from the terminals in one stream along a single link to a multiplexer at the host site which splits the data back into the separate streams from each terminal for presentation to the computer (fig. 2.4).

In this way it appears to each terminal that it has its own point-to-point link to the computer, and each terminal may access the host simultaneously. To achieve this the link between the multiplexers must have sufficient capacity for all terminals to be active at one time. A multiplexer is frequently used to connect many low-speed terminals (eg 300 bps) in an economical manner to a mainframe. Multiplexers are discussed in detail in Chapter 6.

Concentrators

A CONCENTRATOR is a device which can be used as an alternative to a multiplexer to reduce line costs. A concentrator however takes advantage of the situation where terminals are not active all the time and therefore, unlike the multiplexer, not all terminals may access the computer simultaneously. A number of terminals (inputs) are competing or contending for a smaller number of output slots in the same way that extensions on a telephone switchboard contend for the smaller number of exchange lines (fig. 2.5).

A concentrator is a more sophisticated device than a multiplexer and is generally intelligent. It is therefore able to carry out a variety of tasks in addition to concentration (for example, data validation and error checking). Concentrators are discussed in detail in Chapter 6.

Concentrators and multiplexers are used in networks with large

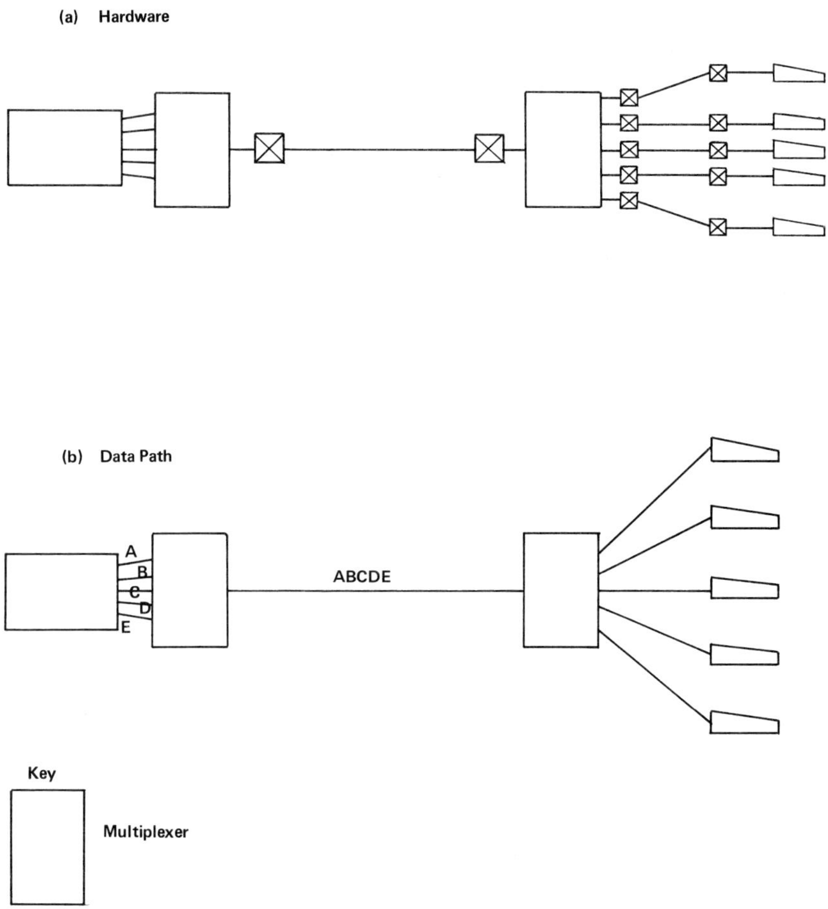

Figure 2.4 Multiplexer link

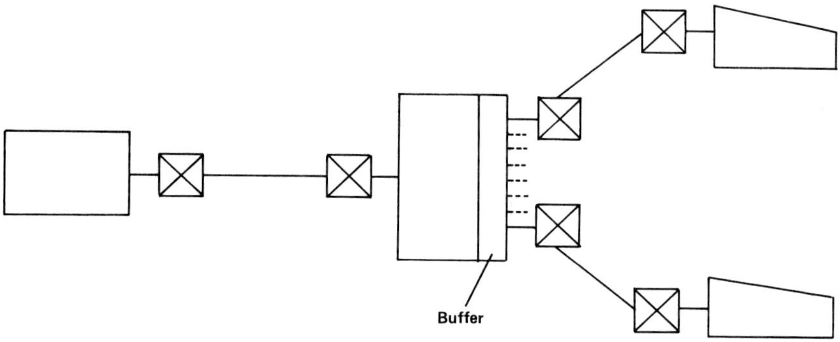

Key

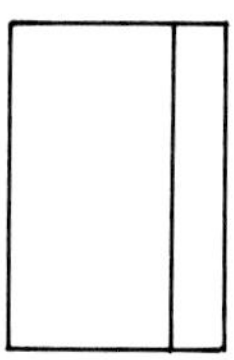

Figure 2.5 Use of a concentrator

numbers of terminals. Generally, in such a case, communications control becomes an extremely onerous task for the host processor. Devices of various levels of sophistication are available to assist in this task. These are known by a variety of names, but COMMUNICATIONS CONTROL UNIT (CCU) and FRONT-END PROCESSOR (FEP) are the most common generic terms. Such a device, placed between the host and the network, reduces the load of communications tasks performed by the host so it is freed for processing tasks. CCUs and FEPs are discussed in detail in Chapter 7.

The British Post Office Communications Services

The Post Office provides a range of services for data transmission, known as Datel Services. A Datel service comprises a circuit on the PSTN or a leased circuit, the associated modem and a full maintenance service. Table 2.1 summarises the currently available Datel services.

Private circuits of varying quality are also available where the Post Office does not provide a full Datel service. Those that are relevant to the data user are narrow voice-band and wideband circuits, as listed below. Tariff S and T circuits do not have maximum speeds at which they can operate; the speed is limited by the modems used.

Tariff H – Telegraph circuit. Suitable for data transmission at signalling rates up to 50 baud.

Tariff J – Telegraphy circuit. Suitable for data transmission at signalling rates up to 110 baud.

Tariff S1 – Basic point-to-point speech circuit. Generally used for data transmission up to 300 bps duplex or up to 600 bps half-duplex.

Tariff S2 – Medium-loss point-to-point speech circuit. Generally used for data transmission up to 1200 bps half-duplex.

Tariff S3 – Low-loss point-to-point speech circuit. Generally used for the data transmission at 1200 bps half-duplex or duplex.

Tariff T – Point-to-point data circuit. Generally used for data transmission up to 9600 bps.

Tariff Y1 – Basically a data circuit designed for up to 12 terminals in a multipoint configuration.

Datel service	Types of circuit	Maximum signalling rate (bps)	Mode of operation	Sync or asynch	Modem type used
200	PC (2-wire)	300	D or HD	asynch	No 2
200	PSTN	300	D or HD	asynch	No 2
600	PC (2-wire)	1200	HD	asynch	No 20
600	PC (4-wire)	1200	D	asynch	No 20
600	PSTN	1200	HD	asynch	No 20
2400	PC (4-wire)	2400	D	synch	No 7
2400 dial up	PSTN	2400	HD	synch	No 7C
2412	PC (4-wire)	2400	D	synch	No 12
2412	PSTN	2400	HD	synch	No 12
48K	PC (4-wire)	50000	D	synch	Nos 8, 9

Key

D Duplex
HD Half-duplex
PC Private circuit

Table 2.1 Datel Services available in the UK

Tariff Y2 – As Y1 but for speeds up to 960 bps.

Wideband circuits at 48 kHz and 240 kHz are also available but can take up to two years to provide.

The Post Office are planning to replace Tariff S1, S2, S3 and T circuits by scheduled circuits A, B, C and D, although no date for this has been announced. These scheduled circuits are similar to the Tariff S and T circuits but some characteristics are more closely specified.

SOFTWARE

Nearly all mainframes on the market today have the hardware capability to handle a data communications system. Software plays an important role in creating an integrated network. This is an area fraught with problems because of the complexity and range of products available, both from the mainframe suppliers and the software houses. In a simple data network the data communications software is generally at the host computer. In more complex networks, containing more than one host or intelligent network elements, such as concentrators and switching nodes, it may be distributed throughout the network.

Figure 2.6 gives one possible breakdown of the software required. In addition to the operating system and applications programs, telecommunications access software is essential. This handles messages, provides message buffering and monitors for failure of devices, lines and of the mainframe. This software can be supplied as a separate software package or as a module of the operating system. It can be located in the mainframe itself or in a front-end processor. Some computer manufacturers produce a range of access methods of different levels of sophistication.

Basically telecommunications access software handles message control in the system. In addition message processing can be optimised within the system by implementing the next level of data communications software – a teleprocessing monitor. This is not however essential.

A teleprocessing monitor makes an on-line system run more efficiently and relieves the user of most or all teleprocessing functions. The programmers are then free to develop applications programs independent of teleprocessing requirements and constraints. Most teleprocessing monitors are limited in the type of devices and in the number of high-level languages they support. Like most software packages, they vary in cost and capability

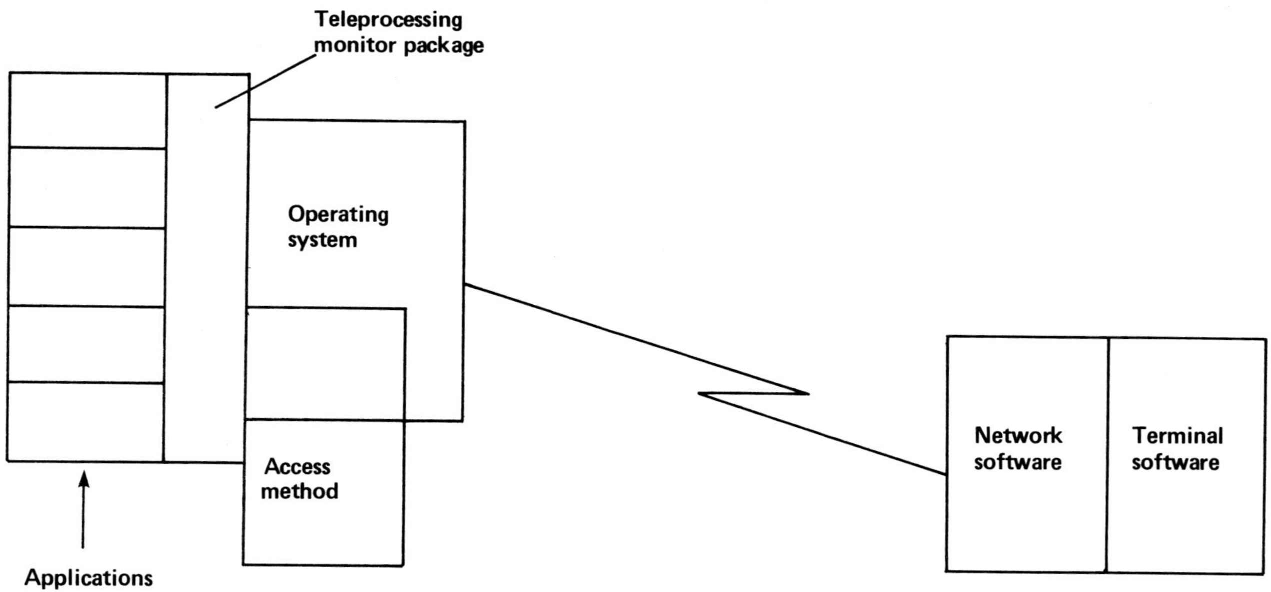

Figure 2.6 Possible software breakdown

from supplier to supplier. There are several NCC publications on teleprocessing monitors:

- *Teleprocessing Monitor Packages for IBM 370* (1978);
- *Teleprocessing Monitor Packages for ICL 2903/04* (1978);
- *Teleprocessing Monitor Packages* (1979).

Many of the major manufacturers have available, or are developing, communications facilities for their products, which enable users to configure, from standard components, networks of a wide range of complexity. These products, called network architectures, are intended to introduce some degree of order into the confused world of telecommunications software.

A 'layered' approach has been adopted by nearly all manufacturers, with functions insulated from each other. Standard interfaces exist between different layers and different functions so that modifications can be made in one layer without affecting another. This allows a user considerable flexibility in expanding a network both in adding more of existing equipment and in adding new equipment and possibly more sophisticated applications. The user however becomes totally dependent on the manufacturer's products as the architectures differ in the facilities which are available, the way in which they are implemented and perhaps most importantly in the jargon used to describe them.

One of the basic differences between network architectures is the type of market at which the product is aimed. The market for communications networks can be broadly divided into two types:

- networks designed to allow terminals to access host mainframes;
- networks which act as distributed processing systems.

Similarly network architectures can be divided into those designed primarily for the first type of network and those for the second. Table 2.2 lists the network architectures of the main computer manufacturers. ICL has not yet announced a formal network architecture although its Basic Mode Protocol permits certain basic hierarchical networks hosted by 1900 and 2900 mainframes to be constructed.

PROTOCOLS

For any exchange of information to be meaningful the parties involved must understand and follow the same set of rules or

Manufacturer	Name of architecture	Bit-oriented line discipline
IBM	System Network Architecture, SNA	SDLC
DEC	Decnet	No
Honeywell	Distributed Systems Environment, DSE	HDLC
Univac	Distributed Communications Architecture, DCA	ADCCP SDLC, HDLC
NCR	Distributed Network Architecture, DNA	NCR/DLC
Hewlett-Packard	Distributed Systems Network, DSN	No
Modcomp	Maxnet	No
Burroughs	BNA	BDLC

Table 2.2 Proprietary network architectures

procedures. For instance in a human conversation, it is understood that only one person speaks at a time, that a common language is used and that any physical signals used mean the same to all involved.

Such rules or PROTOCOLS are as essential in data communications as in any other system for information transfer. Protocols have three essential elements:

- *the message format or syntax.* This specifies the purpose, size and layout of the messages in the protocol, ie provides the form to the sentences of the protocol;
- *the method of procedure or semantics.* This is a list of commands and responses, and their meaning, depending on where in the system they have originated;
- *sequential behaviour procedure.* This element of protocol governs the sequential behaviour as a message's meaning can depend on the message that has preceded it.

Protocol Levels

These are four basic levels of protocols in a network:

- *Physical or electrical interface protocols.* These are the general purpose electrical characteristics for digital and analogue circuits and the procedures used to establish, maintain and disconnect the physical communications link between the DTE and the network or between different DTE units;
- *Link control level protocols.* These protocols govern the exchange of data between the DTE and the network or between the different DTE units;
- *Call control level protocols (network or communications control).* These protocols define the formatting of messages and the control procedures for the transfer of user data through a communications network. This level may appear to perform many of the same functions as the lower levels but the scope is broader;
- *Process-to-process protocols (system control and user communications).* These deal with data exchange on an end-to-end basis, from operating system to operating system or applications program to applications program. They include such processes as file transfer and remote job entry; for instance, an RBT must send information to the applications program in the mainframe in the manner in which it expects to receive

it, eg header of 1 block, 140 blocks of data, end of file marker.

In general, the first two types of protocol above are implemented in hardware while the others are software functions. Figure 2.7 shows where each of these protocol levels sits in a network.

Standards

Manufacturers in the computer industry do not work closely together in software and hardware design. Therefore each manufacturer has developed its own protocols for implementing similar functions. Computer professionals have only to consider the different character sets used in computers and the problems this causes in trying to transfer a magnetic tape, say, from one machine to another with a different character representation, to appreciate the advantages of standardisation. This is especially true in the area of data communications where it is extremely rare for one manufacturer to be able to supply host, terminals and communications hardware.

Many national and international bodies exist which define standard protocols for data communications. Among the most important of these is the Comitè Consultatif International Tèlègraphique et Tèlèphonique (CCITT), a section of the International Telecommunications Union. Also the International Standards Organisation (ISO) and the American National Standards Institute (ANSI) have been extremely active.

Protocols become more complex as the focus shifts from the level of individual pieces of equipment up to the end-to-end level. This increase in complexity is reflected in the extent to which each level has become standardised. To take the two extremes as an example, electrical and physical level interfaces are fairly well standardised, whereas development of higher-level standards is still in the experimental stage.

The most important protocols in the present context are those at the electrical and physical level and at the link level. Electrical and physical interface standards have been developed by a number of different groups. The ISO and CCITT have had responsibility internationally, whilst in the USA the Electronic Industries Association (EIA) and ANSI are the main bodies concerned. The CCITT have produced a series of recommendations, known as the V series, which cover many aspects of data transmission. These are listed in Appendix 4. The most important of these recommendations are:

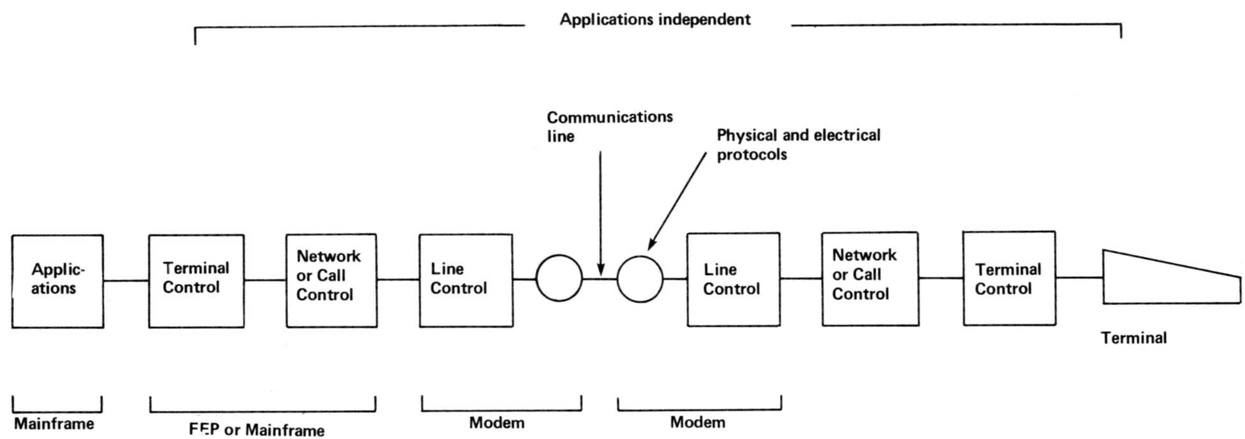

Figure 2.7 Protocol levels in a network

– V24 which defines the functional characteristics of the interface circuits between the modem and terminal up to 20,000 bps;

– V28 which defines the electrical characteristics at the same interface.

These recommendations have been widely accepted by modem manufacturers. The EIA counterpart to V24 and V28 is RS-232-C and these names are often used interchangeably. The V24 circuits are arranged slightly differently to those in RS-232-C. An NCC document 'V24 and RS232-C' is available which explains this difference and its implications. Compatibility can however be easily achieved and RS-232-C and V24 devices can be used interchangeably.

At the link level the influence of computer manufacturers is felt more strongly. The most widely used link protocols today are manufacturers' standards. These include the simple stop-start teletype protocol and the Binary Synchronous Communications Protocol (BSC) originated by IBM in 1968 for use with half-duplex synchronous terminals.

These protocols are character orientated, ie prescribed control characters define information fields and convey logic signalling information, but the trend is now towards bit orientated protocols in which bit strings delimit fields and convey control information. The ISO has developed such a protocol, called High Level Data Link Control (HDLC) for use on full-duplex synchronous links and this is becoming widely used, together with the similar IBM protocol, Synchronous Data Link Control (SDLC).

NETWORK MANAGEMENT

The term *network management* is used here to cover the tasks necessary to keep a network operating at the desired level of performance. Because of the importance to an organisation of the activities generally supported by such a network, maximum availability is essential. Obviously achieving this becomes more difficult and time-consuming the more complex and larger the network. Hence the need for careful consideration to be given to network management as early as the network design stage. This topic is discussed in the NCC publication *Management of On-line Data Communications Systems.*

The principal tasks involved are:

– change control;

- network monitoring and control;
- testing and diagnostic procedures for fault finding;
- fallback procedures.

Change Control

Any change, even if it is only the addition of a single terminal, must be carefully controlled and the implications of such a change on the rest of the network carefully considered.

Network Monitoring and Control

Monitoring and control of a data network may generally include:

- monitoring of the status of lines and individual network elements;
- keeping of a network log for housekeeping purposes;
- provision of a communications interface between the operator and the network;
- provision of supervisory communications between the computer and remote sites.

These activities allow any fault conditions which may be developing, such as deterioration in line quality, to be identified; the statistics gathered may be used in planning future developments to the system. Monitoring may be carried out centrally by the host processor or front-end processor although in a network where intelligence is distributed, concentrators and other intelligent devices may carry out some of the functions. Monitoring may be carried out by continuous automatic on-line recording or periodic off-line checks, the level of sophistication depending on the sophistication of the network software.

Testing and Diagnostic Procedures

Testing and diagnostic procedures must be defined so that any fault in the network may be located and repaired expeditiously. These procedures will vary in complexity and depend primarily on three factors:

- the number of suppliers involved in the network;
- the degree of testing that the organisation is prepared to carry out itself before calling in the suppliers;
- the degree of fallback facilities provided (discussed below).

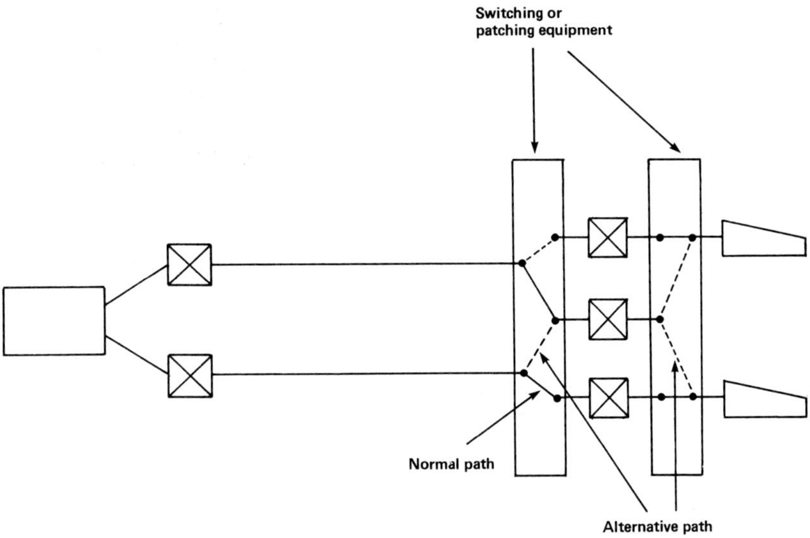

Figure 2.8 Use of Switching and Patching equipment

In order to carry out fault-finding certain items of equipment are necessary. This TESTING AND DIAGNOSTIC EQUIPMENT is described in detail in Chapter 9. The amount of equipment required depends on the amount of testing the network operator is prepared to undertake, for a simple network a few inexpensive items are generally all that is required.

Fallback Procedures

When a fault occurs in a network, the network operator must decide whether an alternative means of service or 'fallback' service is to be provided to users. This will depend on the importance of the applications using the network to the organisations and may vary from link to link. Fallback facilities may be provided in several ways:

- manual processes and delivery services;
- rerouteing the data through an alternative path in the network;
- use of standby equipment.

To carry out the second two alternatives equipment known as SWITCHING OR PATCHING EQUIPMENT can be used. This is placed between various devices in the network, most commonly, the computer and modem, and modem and transmission line. This allows spare equipment to be switched into the network or for the network to be reconfigured (fig. 2.8) in an easy and neat manner. In addition it facilitates an important testing procedure known as 'a loopback test'.

Therefore, while switching and patching equipment is by no means essential, it is extremely useful at certain points in any network. The equipment is discussed in detail in Chapter 8.

3 Considerations for Simple Network Design

INTRODUCTION

The use of data communications is becoming increasingly common and even small organisations are now able to consider implementing some communications links for remote terminal access. This chapter outlines the considerations which must be borne in mind when designing a simple network. The aim is a satisfactory end result for both the user and the data processing department (usually responsible for the network's technical performance).

It is assumed that the organisation already has its own in-house computer system, and the reasoning behind the decision to extend its capability to include on-line applications is not examined. Such reasons are peculiar to each individual organisation. However, the steps which must be followed to carry the decision through to implementation are the same. (There are many good books available on the subject; for example, *Planning for Data Communications* by Bingham and Davies (MacMillan Press 1977) and *Systems Analysis for Data Transmission* by James Martin (Prentice Hall 1972).)

The selection of hardware for the network is only one step in this process but one on which much depends. The aim of this chapter is to fit the hardware selection exercise into its place in the whole process, and to highlight the many factors which can influence the final selection. Such factors can frequently result in compromise being necessary between the 'best' piece of hardware and that which fits most appropriately into the overall design considerations.

The important steps in the process (fig. 3.1) are discussed briefly in the remaining sections of this chapter. Parts of the process are iterative while other steps can be carried out in parallel. The whole process is generally carried out within budgetary constraints and, just as important, time constraints.

It is essential to start the planning process early and to allow a realistic timescale for design and implementation of the network. A technique such as critical path analysis can be extremely useful in planning a network and in facilitating adherence to the planned schedule. Hardware supply and software modifications can often take longer than expected but early planning can help prevent costly delays.

IDENTIFYING USER NEEDS

The first step in implementing a computer network is for the computer department and the application users to investigate together the feasibility of the scheme. This requires a detailed investigation of the users' requirements so that these are taken into consideration in the technical solution selected.

Such a study is generally carried out by a systems analyst from the data processing department, with considerable involvement from the user departments.

Figure 3.1 Steps in the design of a data communications network

* Study user needs to identify
  - location of terminals
  - location of hosts
  - traffic volumes and frequency
  - types of messages
  - types of application
  - features required of terminals
  - acceptable downtime
* Decide on whether existing mainframe supplier will supply terminals, front-end processor (if used) and software.
* Consider software changes necessary, eg is the operating system suitable, teleprocessing monitor, protocols, etc.
* Draw up a list of terminal characteristics and select possible terminal types.
* Draw up network configuration and select circuits.
* Consider reliability and network management needs.
* Select hardware.
* Put it all together and make sure it works.

The ultimate success of the system depends upon its acceptability to all users. Most systems have more than one type of user, and the requirements of each group can frequently be different. The possible user types include business users, computer operations staff, data entry personnel, irregular users such as managers requiring statistical information on an ad hoc basis, and staff who never physically come into contact with the system but who supply data for entry and then use the output.

The important steps in this feasibility study are:

- estimation of traffic volumes and frequency;
- identifying of location of terminals;
- identifying of location of host;
- response time requirements;
- identification of message types and of interactive and batch applications;
- identification of the features required of terminals;
- estimation of acceptable downtime.

Each step is considered briefly below with the important points highlighted.

Estimation of Traffic Volumes and Frequency of Use

The basic parameters for the design of a data communications system are an estimate of the traffic it will have to carry and the frequency of use. Both average and peak values are essential.

For an existing manual system which is to be transferred to the network such figures will either already be recorded or can be obtained quickly by a simple data collection exercise. For a new application some estimation will be necessary. It is important that this is done as realistically as possible, but that the numbers should be over- rather than under-estimates. It will not affect users if the final system has a degree of spare capacity but users will be subjected to an inferior service if the system is overloaded.

Indeed, the system must be designed around peak traffic volumes so that under normal operating conditions there will be ample spare capacity. It is also essential to try to anticipate any likely increase in traffic in the future so that the system will not have to be modified due to overloading soon after installation.

It is necessary to estimate, for each user, the frequency of system use. This is needed in the calculation of the number and location of terminals required. Values for average daily use, peak daily use and at what time this occurs, average monthly use, and peak annual use should be identified for each user and each application.

Identification of Location of Terminals

To decide on the location of terminals it is necessary to know the location of every possible user of the system. This information will have been gathered when the frequency of use of the system was analysed. It will frequently not be necessary for every user to have a terminal but care must be taken that terminals are located sufficiently close to users otherwise they will be reluctant to use the system because of the inconvenience.

Such decisions will initially be based on frequency of use and distance. There may also be organisational reasons which dictate the final number and location of terminals.

Identification of Location of the Host

The organisation must have established the location of the host of the data communications system before any configuration can be designed. When installing a network however it is not uncommon to consider upgrading the existing computer, acquiring a new computer or acquiring an additional one to complement the existing equipment.

It may also be to the organisation's advantage, if a new or additional computer is to be obtained, to consider relocating the computer centre so that it is in the optimal position from the point of view of the data communications system, ie the site at which line costs would be at a minimum. In a country the size of the UK however, where line lengths can never be excessive, the cost of relocation may well be greater than any savings gained.

Response Time Requirements

Response time in an on-line system is strictly defined as the interval of time that elapses between the completion of the user's last entry at the terminal and the display at the terminal of the first character of output received from the computer system or network, in response to the command just initiated. This output does not necessarily have to be the reply to the user's entry. An acknowledgement that the entry has been received will often be sufficient.

Such a time interval is made up of several independent elements, the most important of which are:

- the time taken for the two-way communication itself;
- the availability of processing time at the mainframe;
- the data retrieval process.

Each of these elements must be separately assessed and the design of the system be such that the actual response time requirements of users can be met. Where the network is a simple point-to-point network the communications links do not make a significant contribution to the response time but in a complex network of widely scattered terminals they can become responsible for the longest delay.

The cost of a fast response time can be high but as it is not necessary for all response times in the system to be the same it is possible to tailor the response time to the functions the system performed. Users of an enquiry-response application for example require a faster response than those operating an RBT; generally the greater the interaction between the user and the terminal the faster the response necessary from the system for the user to operate effectively.

Response times are not discrete elements but distributions, and it is not sufficient to design the system with only a satisfactory average response time. The spread of the distribution must be narrow enough to prevent a high percentage of the response times being much greater. The standard deviation is as important as the average value. Bearing in mind the user's requirements and the application, the systems analyst must assess a response time that satisfies both and is also not prohibitively expensive to achieve.

Identification of Required Type of Data Communications

In order of increasing degree of complexity the available types of on-line data communications can be classified as:

- remote batch;
- on-line data collection;
- enquiry-response system;
- real-time system.

A detailed description of these types can be found in *Planning for Data Communications* (Bingham and Davies, MacMillan Press, 1977). Each demands different features from the data communications system, such as line speeds, protocols, and complexity of software in the host. It is the characteristics of the application itself which ultimately enable the systems analyst to decide which is the most suitable. If more than one application is involved, several types may be used on the same data communications network. Several factors are important in making this decision; namely, the message length of the application, the response time required by the users and the importance of up-to-the-minute information. It is, for example, unnecessary to install a real-time payroll system with hours worked by staff entered daily, if the information is required to be up-to-date only once a week or once a month. In such a case a remote batch system will be quite adequate.

Identification of Features Required by the Terminals

Before approaching terminal manufacturers the data processing manager must have a detailed specification of the terminal type required. Such a specification should include the following:

- What type of data communications system is the terminal to be used for?
- What environment is the terminal to operate in? This indicates the degree of robustness required by the terminal and may also influence its size.
- What method of input is required? This will generally be keyboard, punched card, badge reader, magnetic tape or paper tape.
- What method of output is required? It is especially important here to identify if users require hard copy output or not.
- What character set is required? Some terminals with keyboards do not have the full range of characters, for example £ or $ are not always present. Also if a large amount of the data entered via a keyboard is numeric it is possible to provide a keyboard which has a separate numeric pad.
- What degree of error control and correction facilities are required? This influences the decision on whether the terminal is to be synchronous or asynchronous.

Acceptable Downtime

The prospective users of the system must be consulted on the length of time for which a failure of the system is tolerable.

Different locations carrying out different applications or different aspects of the same application may not have the same requirements. The critical locations must be identified and this information used at a later stage in planning fallback equipment and procedures.

Other Information

Whilst carrying out the feasibility study the system analyst should gather as much additional information on the users of the new system as possible. This will be useful in the later stages of design and implementation. For example, note should be taken of the potential abilities of different types of user to use the system effectively when it becomes operational. This will be used to assess training requirements. The requirements for security of access to the system must also be considered so that suitable safeguards can be developed.

SINGLE OR MULTISUPPLIER

As already mentioned, it is common for an organisation to consider the installation of a data communications system at the same time as its mainframe is being upgraded or a new or additional mainframe is being acquired. This brings the mainframe supplier into the process at an early stage. Whether such a change is taking place or not, the data processing manager must consider the role of the mainframe supplier in the data communications system early in the planning stage.

Most mainframe suppliers do not include data communications hardware such as modems, multiplexers, test equipment or switching equipment in their product ranges. It is in the area of terminals, concentrators, front-end processors and software that their involvement is possible. There are obvious advantages for an organisation inexperienced in data communications to go to their existing supplier for these products as:

- there is less chance of problems caused by hardware and software incompatibility;
- the organisation already has confidence and a good working relationship with its supplier;
- maintenance and fault correction is made easier. The fewer suppliers involved in the network the fewer problems that are generally encountered in this area.

Despite this, the data processing manager must be prepared to consider alternative suppliers, especially if there is a requirement for any special features (for example, in terminals). Staying with the mainframe supplier can be an expensive way of ensuring peace of mind. The chief danger lies in being 'locked in' to a particular manufacturer by adopting for instance a protocol, language or hardware interface in such a way that it would not be cost-effective to introduce another manufacturer's equipment in the future when requirements may well have changed.

SOFTWARE CONSIDERATIONS

Software plays an important role in a data communications network. Additional software for controlling the network will be essential and the data processing manager should approach the mainframe manufacturer to find out what is required for the particular system.

A considerable amount of effort will also be necessary to convert any existing applications programs and write new applications to interface with the network. It is often sensible to undertake this work in a phased approach. This not only helps if there are resource constraints but allows the new systems to be absorbed gradually into the organisation, allowing problems in one application to be cleared up before another is added to the network.

TERMINAL SELECTION

Terminals on the market today fall into four categories: teleprinter terminals, VDU terminals, batch terminals and intelligent terminals. By this stage in the planning process the decision will have been made as to which of these types is required. In most cases a mixture will be used depending on the requirements of a particular application.

To most users of the network the terminal is the only part with which they will become familiar. Consequently the acquisition of terminals becomes extremely important: they must satisfy technical and economic criteria, and also please the user.

Using as a basis the information collected during the assessment of user needs, a list of criteria can be drawn up with which to assess suitable terminals. The selection process can then proceed in a systematic manner using much the same approach as that described in the remainder of the book for the selection of other items of data communications hardware. The enormous range of terminals available and the wide variety of features offered make a description of the selection process more lengthy however. The NCC publication, *Selection and Use of Terminals in On-Line Systems*, gives an excellent description of the process.

NETWORK CONFIGURATION

By this stage the data processing manager will be able to draw up a diagram showing the location, type and probable model of each terminal, the location of the mainframe, the response time requirements, and the traffic flows between all sites. The manager must now decide how to configure the network. The major decisions to be made are:

- Are the communications links to be PSTN or leased circuits?
- What speeds are the links to be?
- Are links to be used for half-duplex or duplex transmission, and what protocol is to be used?
- What is the physical arrangement of the links?

The basic objectives in determining the network configuration should be to minimise the total communications link distance, and therefore its cost; and to equalise as far as possible the workload over available links and equipment.

PSTN or Leased Circuits

The British Post Office will supply private or PSTN circuits at various speeds for data transmission which takes place between the different sites of an organisation. It is possible to configure a network with either PSTN or leased circuits, or with a mixture of both, as well as with in-house circuits installed by the organisation itself.

Leased circuits of varying quality are available. Generally the normal quality circuits, similar to those used on the PSTN are adequate for data transmission at low and medium speeds. At speeds of 4800 bps or above, such circuits, as well as the PSTN, are not adequate and Tariff T leased circuits are recommended.

A leased circuit is a permanent link between two points; a dial-up link is more flexible, and ideal in situations where a terminal may wish to access more than one computer on different occasions. Additionally, if a PSTN call is broken, the user has only to hang up and redial to obtain a new circuit. If a private circuit fails, the user has no alternative route unless there are arrangements for a back-up leased link or for a PSTN back-up arrangement.

PSTN line charges are usage dependent with a small monthly rental while private lines have a usage independent charge in the form of a high fixed monthly rental. Both charges are however distance-dependent. This tariff structure means that for low volumes of data the PSTN is the more economic while for high volumes a leased line is the more economic alternative.

It is possible to calculate at what volume of traffic a private line becomes the cheaper alternative (fig. 3.2). It is usually this difference in pricing structure, and consequently different cost to the user, that dictates whether a line should be leased or PSTN.

Assumptions

1 Work is done during office hours, half before 1 pm and half after 1 pm.

2 Installation costs are not included

3 The calculation is based on connect time only and not on the volume of data actually transmitted.

Break-even points are at A, B, C, D, E, and F (see fig. 3.2).

For example at point C, if the connect time is greater than 12 hours per month, it is cheaper to use a private line at 300 bps.

Speed

The required speed for each line must be calculated using the anticipated volume of traffic, the user's response time requirements and the type of terminal to be used. For instance, there is little point in using a 1200 bps link for a teletype which is limited by the typing speed of its user for input and by its own printing speed of 30 characters per second for output. In general, such terminals would require only 300 bps links while remote batch terminals (RBTs) and visual display units (VDUs) would require higher speeds.

If a multipoint line is to be used, the calculation of the required speed is more complex but the total volume of data to and from all devices on the line must be taken into consideration.

Full- or Half-Duplex

The choice of full- or half-duplex working is influenced by the

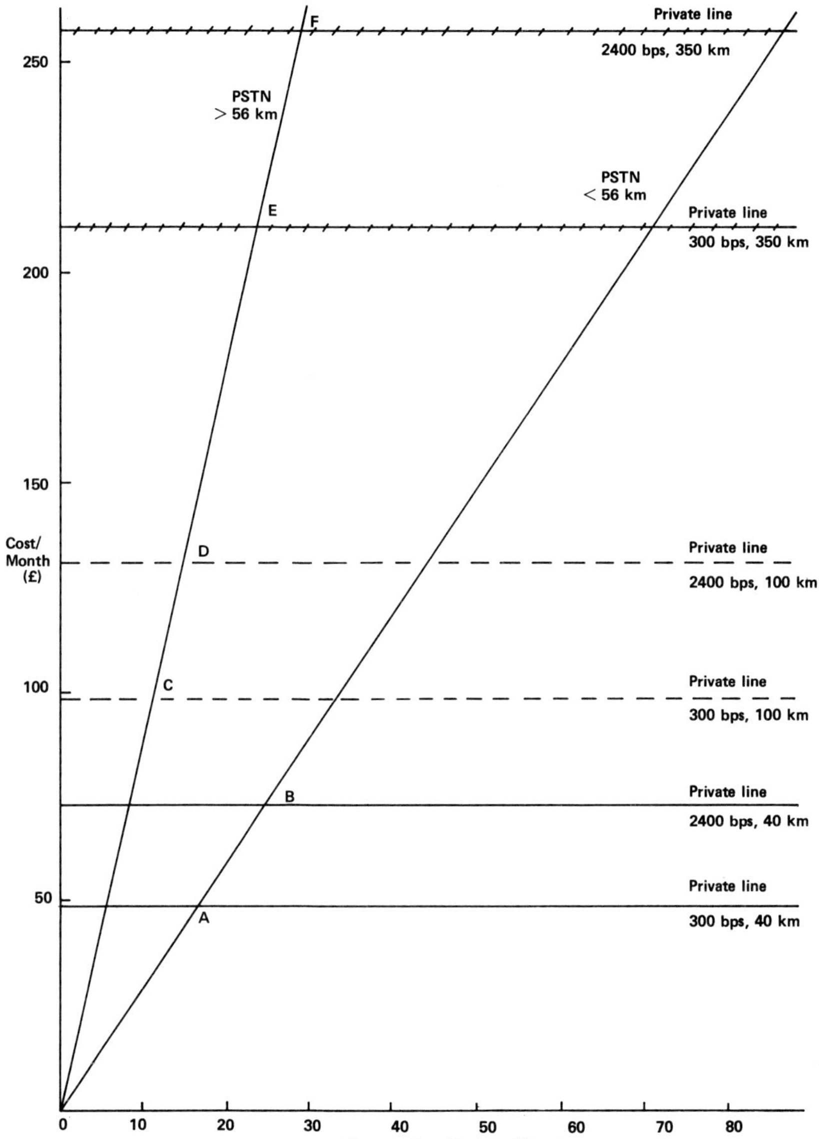
Private line
2400 bps, 350 km
F
PSTN
> 56 km
PSTN
< 56 km
E
Private line
300 bps, 350 km
D
Private line
2400 bps, 100 km
C
Private line
300 bps, 100 km
B
Private line
2400 bps, 40 km
A
Private line
300 bps, 40 km
Cost/
Month
(£)
250
200
150
100
50
0
10
20
30
40
50
60
70
80
Connect time (hrs/month)
Figure 3.2

response time required, by the characteristics of the terminals and by the line protocol likely to be used. It is common for networks which do not actually require data to be transmitted in both directions simultaneously to use full-duplex data channels so that the turn-around time of the channel does not lead to an unacceptably long response time. Turning the line around requires special facilities in the computer software and terminal and computer interface hardware and this in itself can be a reason to avoid half-duplex working. At data rates up to 1200 bps it is possible to obtain full-duplex working on a two-wire half-duplex connection by using one pair of frequencies for transmitting and a separate pair for receiving. Such an arrangement is becoming increasingly common and the maximum speed of operation is becoming higher as modem technology improves.

Physical Arrangement of the Links

The alternative ways of connecting a terminal to a host mainframe are:

- a point-to-point link;
- a multipoint link;
- a multiplexer;
- a concentrator.

The use of point-to-point links, whether PSTN or leased, will result in a higher total link cost (costs are proportional to distance), and if each terminal is not heavily utilised it can be a very inefficient use of resources. Each line requires a port at the mainframe: provision of sufficient ports, if there are many terminals on the network, can cause problems. A point-to-point network is the easiest type to design, implement and maintain. It is especially useful in cases where there are terminals at sites which are themselves widely scattered but all within a fairly short radius of the host computer.

A multipoint link can increase the utilisation of transmission resources by permitting more than one terminal to share a link, although only one may actually transmit at any time. Use of such lines requires more intelligence and this increases the costs. This must be balanced against the savings in line costs. Multipoint lines must be leased: the arrangement cannot be used on the PSTN. There are no rules for deciding how to connect terminals together via a multipoint link. In cases where point-to-point routes from the terminals to the host partially overlap, it can be obvious; in other cases not so. The Post Office will however generally offer suggestions on the best way to configure a multipoint link.

There is a definite probability that out of a number of terminals on one multipoint line, more than one will wish to use the line at the same time. Knowing the volume of traffic and frequency of use of each terminal it is possible to calculate the probability of this happening in a given time period. In this way the number of terminals on the multipoint line which gives an acceptable level of service to users can be calculated. In practise it is quite common to have at least four or five terminals on one multipoint line, the Post Office permit up to twelve branches.

Use of a multiplexer or concentrator can reduce line costs without requiring any additional ability at the terminal. In applications where it is required to transmit data from a number of low-speed terminals at the same or at geographically convenient sites, savings in line costs can be achieved, especially over long distances. Figure 3.3 shows a typical calculation to see if it is cheaper to use a multiplexer than several point-to-point links or a multipoint link. A cost that is frequently forgotten in this calculation is that of the two high-speed modems required. A frequency division multiplexer can also be used in a manner analogous to a multidrop link. This can be a very cost-effective arrangement and is described further in Chapter 6.

Concentrators can also have the effect of distributing intelligence throughout the network, thus reducing the total volume of data transmitted. For instance editing or formatting of data can be done at the concentrator and only correct data need be transmitted to the host mainframe.

The multipoint, multiplexer and concentrator solutions do however lower the reliability of the network since in each case more than one terminal is dependent on one link. This factor alone can be sufficient for some organisations to reject these solutions even if they are more economical than a point-to-point network.

Consider a simple example of 6 teletypewriter-compatible terminals located one at each of 6 sites. The sites are close together, and approximately 300 kilometers from the central computer site. The sites are approximately 10 kilometers apart and must be connected to the central site by 300 bps leased circuits.

A. Point-to-point links

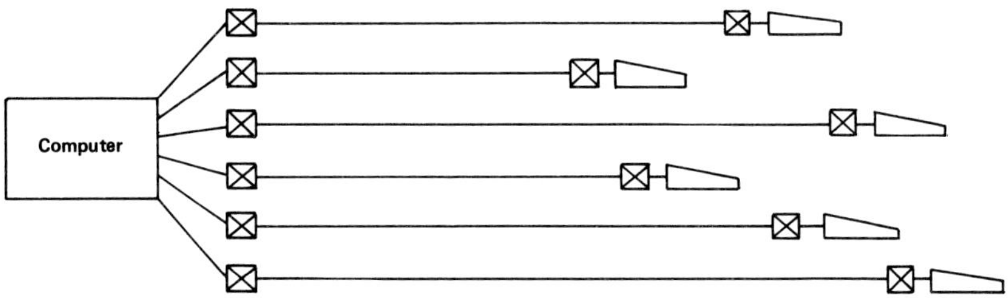

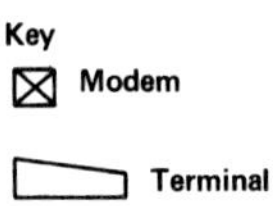

Connection charges			
6	300 bps tariff S1 lines (300Kms)	£ 450	
12	Modem 2s from the Post Office	£1200	£1650
Rentals per month			
6	300 bps tariff S1 lines (300 Kms)	£1117.5	
12	Modem 2s from the Post Office	£ 24.96	£1142.46

Figure 3.3 Comparison of costs of using point-to-point, multipoint or multiplexer configurations

(Fig. 3.3 cont'd)

B. Multipoint links

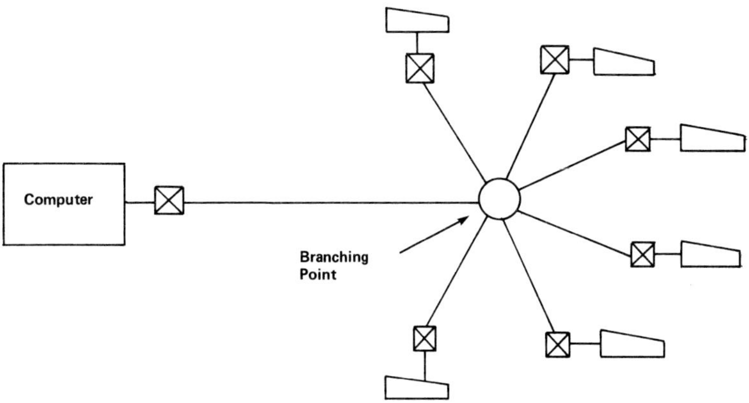

Connection charges

1	300 bps tariff Y1 (300 Kms)	£162.2	
7	Modem 2s from Post Office	£700	£862.2

Rentals per month

1	300 bps tariff Y1 line (300 Kms)	£217.08	
7	Modem 2s from the Post Office	£ 14.56	
1	branching point	£ 8.33	
6	local ends	£45	£284.97

C. Multiplexer link

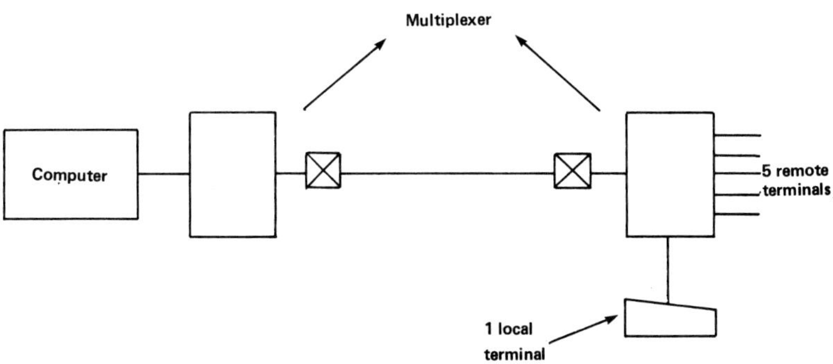

Connection charges

5	300 bps tariff S1 lines (10 Kms)	£125	
10	Modem 2s from the Post Office	£1000	
2	Modem 7s from the Post Office	£200	
1	2400 bps tariff T line (300 Kms)	£120	£1445

Rental charges

5	300 bps tariff S1 lines (10 Kms)	£80	
10	Modem 2s from the Post Office	£20.8	
2	Modem 7s from the Post Office	£55	
1	2400 bps tariff T line (300 Kms)	£257.5	£413.3

A simple multiplexer with 6 asynchronous input channels costs typically £4000 – £4500.

Writing off the cost of 2 (at £4000) over 4 years will add £165 to the monthly rental.

Summary table

	Point-to-point	Multipoint	Multiplexer
Connection charges	£ 1650	£862.2	£1445
Monthly costs	£1142.46	£285	£578.3

All costs are exclusive of VAT

Maximum connection charges are assumed for modems

RELIABILITY, SECURITY AND RECOVERY

Increasingly companies are depending more upon their on-line computing systems and systems designers must therefore take considerable care to build a reliable and secure system with adequate backup (or fallback) provisions.

The risks which must be assessed in this area may be summarised as:

- hardware failure;
- communication link failure;
- system software failure;
- applications software failure;
- power supply or air conditioning failure;
- human error;
- fire;
- flood;
- theft and fraud;
- malicious damage and sabotage.

Once these risks have been assessed, suitable procedures can be taken to minimise the danger.

For each part of the network it is necessary to assess the degree of fallback facilities required both in links and hardware, and how much money you are prepared to invest in achieving this. A common arrangement is to use dial-up circuits as a fallback arrangement if a leased line falls. This may operate at a lower speed but the data is still able to get through. This fallback arrangement must be foreseen in the planning stage as it requires special features on the modems used.

If a link is extremely important it may well be worth the expense of having a spare leased circuit but this arrangement can often not be as foolproof as it seems. If, for example, both leased circuits follow the same route, a failure in the exchange at either end or en route could result in the loss of both links.

An alternative arrangement, possible in some shaped networks, is to make arrangements to transmit data via a different route if part of the network fails. Consider the network shown in Figure 3.4. If the link between A and C should fail then it might be more satisfactory to use the route A-B-C as a fallback arrangement; the

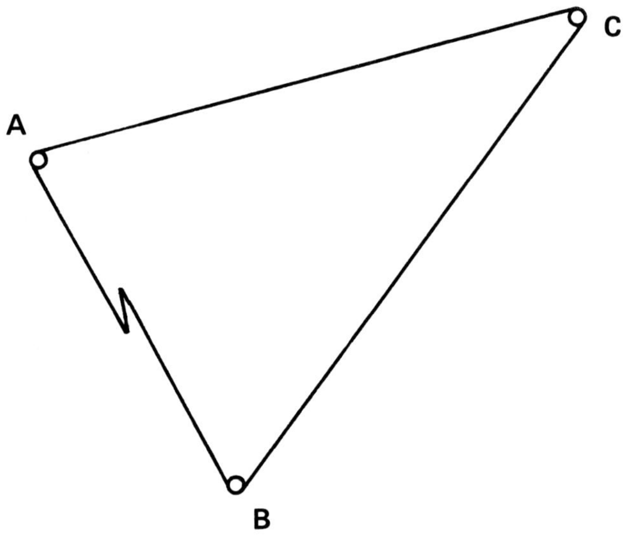

Figure 3.4 Alternative routeing

connection from A to C being dial-up or a standby leased line.

At vital points in the network it may be necessary to provide duplicate hardware modules. More generally however failure of a link for a matter of hours is not catastrophic and provided repair is effected quickly standby hardware is not necessary. At a site with many modems it is worth having one standby for about four or five operating modems as this will offer a cheap method of fall-back.

By bearing in mind the need for reliability, security and recovery at an early stage, many problems can be foreseen and catered for in the network design. Once the network is operational from a hardware point of view, a certain degree of self-help, a good relationship with suppliers and good maintenance contracts should help to ensure high availability of the network.

SELECTION OF HARDWARE

The hardware selection process must be undertaken in a systematic manner so that a quantitative comparative analysis can be made. The stages involved are:

- contacting the Post Office for advice and quotations on links required and on suitability of sites in the network;
- drawing up of a list of criteria for each item of equipment and assigning rankings to the criteria;
- establishing a scoring system for each criterion;
- identifying suppliers who are to be invited to tender;
- preparing and issuing of invitations to tender;
- evaluating each manufacturer's equipment against this scoring system;
- evaluating each manufacturer's contribution to the range of equipment required to decide if it is possible to use only one supplier;
- final selection of each item of equipment.

These stages are described in greater detail in the remainder of the book.

IMPLEMENTATION

Once the data communications system has been planned and all the necessary hardware and software selected, the data processing manager will be able to draw up a detailed implementation plan. As this will inevitably involve the co-ordination of several suppliers it is important to get them to agree to realistic delivery and installation dates. For example, there is no point getting the hardware delivered within 3 months if the Post Office is unable to supply the necessary lines in less than a year (a common enough occurrence). A delivery point for each item must be identified at an early stage. Is the hardware to be delivered immediately to its final site or is everything to be delivered centrally so the network can first be configured under one roof and tested?

During the elapsed time between ordering and delivery, several tasks must be carried out, notably the preparation of sites for the installation of the equipment. In the simplest case this may just be ensuring that there is a suitable power supply nearby to a modem's location, whereas preparing the site for a complex front-end processor could require considerable work. Training personnel can also be planned during this time.

If the user is making additions to an existing data communications system rather than installing a new network the problems are somewhat different. Care must be taken to install the new equipment without too much disruption to existing service.

Each part of the network will require acceptance testing after its installation in addition to a final acceptance test of the whole system. These tests will of course be carried out with the suppliers present. For the final tests to be as meaningful as possible some of the application programs for which the system is designed should be used. The system should not be made available to users, even for training, until it is thoroughly tested and performing to specification. Otherwise confidence in the system will not develop and its eventual acceptance to users will be delayed.

Once the system is installed satisfactorily the user must cope, should a failure occur. A sensible user will have a comprehensive maintenance agreement with the suppliers but this still leaves the responsibility of identifying the faulty piece of equipment so that the correct supplier can be called in. Therefore one person in the organisation must have received sufficient training in data communications principles and hardware to enable an effective search for faults to be carried out.

This training should preferably have been given before installation so that the person can gain practical experience during the installation phase. The job itself will generally not occupy the person full-time but when a problem occurs priority should be given to fault-finding in the data communications network. A typical job specification is given in *Fault Diagnosis in Data Communications.*

4 Criteria for Selection of Hardware

INTRODUCTION

In selecting any product one has to draw up a list of criteria by which it will be judged and evaluate each model by how well it satisfies this list. In the selection of data communications hardware, some rigour is required to ensure that the large sums of money involved are spent as wisely as possible. A rigorous evaluation method also allows this selection exercise to be carried out successfully by a person with little previous experience. This chapter describes an evaluation method which can be used for all items of data communications hardware.

The first step in the evaluation process is the establishment of the list of criteria. Some of these criteria, which we will call general criteria, will be important for all the items of hardware discussed in this book. Examples of this category include cost and reliability; a complete list is given in Figure 4.1 and they are discussed in detail below.

There is however a second group of criteria which is unique to each item of hardware; for instance, the standard features looked for in a front-end processor will differ considerably from those required of a multiplexer. These specific criteria are mentioned briefly below but are covered in depth for each item of hardware in Chapters 5, 6, 7, 8 and 9.

THE EVALUATION PROCESS

Selection of any item can easily be influenced by many factors which bear no relationship to its actual suitability. For instance, a person may reject one manufacturer's product because the salesman is unacceptable; or may buy another's because of the

Figure 4.1 Considerations for general criteria

* Delivery time
* Simplicity of installation
* Reliability
* Maintenance support
* Post Office regulations
* Compatibility and standards requirements
* Supplier factors
* Costs

manufacturer's advertising campaign, regardless of the products' relative merits. This subjective approach can result in costly mistakes and must be avoided in the selection of data communications hardware: the success or failure of the system can have an enormous impact on the organisation's business. It is therefore essential to carry out a more rigorous and more objective analysis.

There are many methods of analysis, some based on a points scoring system and others on a financial scoring system. One of the simplest and most effective methods of analysis is the 'weighted ranking' analysis whereby each item is marked against a list of predefined criteria which have been weighted according to their importance. This is done on a points basis. The approach is straightforward to understand and implement and can be applied equally effectively to all the items of data communications hardware discussed in this book.

The steps in the analysis are:

i) draw up a list of criteria on which the evaluation is to be based;

ii) identify essential criteria;

iii) assign weightings to the remaining criteria;

iv) draw up a list of factors for awarding scores to each criterion;

v) evaluate each model;

vi) compare results and make selection;

Steps (i) to (iv) are best carried out by a committee of two or three rather than an individual. This is because of the subjective

Plan	Service provided		Maintenance cost for a purchased modem 2200/24	Maintenance cost for a leased modem 2200/24
	Hours	Days		
1	0830 – 1600	Mon-Fri ex. ph	£42	N/A
2	0830 – 1600	Sun–Sat inc. ph	£92	£50
3	0600 – 2150	Mon-Fri ex. ph	£92	£50
4	0600 – 2150	Sun–Sat inc.ph	£142	£100
5	0001 – 2400	Mon-Fri ex. ph	£142	£100
6	0001 – 2400	Sun-Sat inc.ph	£192	£150

Notes:
1. ph = public holiday
2. costs are per annum and are exclusive of VAT
3. the service engineer will normally arrive within 4 hours of the request for assistance

Table 4.1 Example of maintenance contracts available on data communications hardware from a major supplier

nature of the tasks; a committee is more likely to produce an unbiased view than an individual. To ensure that the interests of all parties are taken into consideration the committee should include:

- a senior member of the data processing department to represent the technical viewpoint;
- the systems analyst, responsible for the feasibility study, who should therefore be able to understand both the technical and user viewpoints;
- the person who will ultimately be responsible for network management and fault-finding.

Step (i): Draw Up a List of Criteria

The first step is to draw up a list of criteria which are important in the selection of the piece of equipment. Some of these (for example, cost and compatibility) will feature in the lists for all types of hardware required for a data communications network; other more specific features will be unique to each type of hardware.

It is important to ensure that each criterion is capable of qualitative analysis and therefore clearly defined. For example, reliability is too vague a term for a criterion as there are many ways in which reliability can be assessed. Reliability, mean time between failure, is a far more explicit criterion and can be evaluated. The committee must also be aware of criteria which are implicit in others, so that each factor is not evaluated more than once.

In this chapter and in Chapters 5 to 9 a complete list of criteria will be drawn up for each item of data communications hardware discussed. A user of the book will then only have to decide on their relative importance to his organisation. This might well result in some being excluded from the list.

Step (ii): Identify the Essential Criteria

Once the list of criteria has been drawn up, they must be ranked according to their importance, but firstly all essential criteria must be identified. If a piece of hardware does not satisfy one of these essential criteria then it is totally unsuitable and need not be considered further. The list of essential criteria may be used to draw up a shortlist of suitable items which can then be evaluated further under the remaining desirable, but not essential, criteria.

In the evaluation of data communications hardware, some criteria will always be essential: for instance, a modem must always have Post Office approval. Others may well be essential in some circumstances but not others; for instance, an essential feature for a multiplexer needed to multiplex asynchronous teletypes is that it can handle asynchronous devices, but a multiplexer used in different circumstances may not need this capability.

Step (iii): Assign Weightings to the Criteria

Once the essential criteria have been identified, a weighting matrix can be constructed so that each of the remaining criteria can be ranked against every other. The following scoring system should be used:

Criterion A more important than B:	Score 1
Criterion A equally important as B:	Score 2
Criterion A less important than B:	Score 3

The whole matrix, except the diagonal, should be completed, even though either side of the diagonal will be complementary. Complete the matrix by comparing the criteria along the horizontal against those along the vertical and totalling downwards. In this way the higher the weighting the greater its importance. Therefore the equipment selected will be the one with the highest final score.

Consider the example of selecting a book to read on a journey. Simply, the desirable criteria by which it may be judged are:

- its price;
- its length;
- its subject matter;
- its author.

Figure 4.2 shows these set out in a weighting matrix and ranked against each other. For instance entry A is price compared with author and the score assigned is 1, ie price is of less importance than author. Entry B, author compared with price, is naturally assigned the score 3 as it follows that the author is more important than the price.

An essential criterion in this example could be that I have only £1 in my pocket and therefore the book must cost less than £1. This does not prevent the price of the book being a desirable

		Price	Length	Subject	Author	
	Price		2	3	③	entry B
	Length	2		3	2	
	Subject	1	1		2	
entry A	Author	①	2	2		
	Total	4	5	8	7	

Normalising the weightings to the scale 1 to 10

i) Price weighting

$$= \frac{10}{2\,(4-1)+1}(4-4+2)$$

$$= \frac{10}{7} \times 2 = 2.85$$

$\therefore$ Price weighting = 3

ii) Length weighting

$$= \frac{10}{2\,(4-1)+1}(5-4+2)$$

$$= \frac{10}{7} \times 3 = 4.28$$

$\therefore$ Price weighting = 4

iii) Subject weighting

$$= \frac{10}{2\,(4-1+1}(8-4+2)$$

$$= \frac{10}{7} \times 6 = 8.57$$

$\therefore$ Subject weighting = 9

iv) Author weighting

$$= \frac{10}{2\,(4-1)+1}(7-4+2)$$

$$= \frac{10}{7} \times 5 = 7.14$$

$\therefore$ Author weighting = 7

Figure 4.2

criterion, as being naturally frugal, I can still be interested in getting a cheap book *below* the upper limit of £1.

Once the matrix has been completed and the scores totalled downwards, these scores can be used as weightings. It is more convenient however, especially if there are many criteria, to convert these scores to weightings in the range 1 to 10. To do this the following formula must be applied:

If criteria A has score X then its weighting will be calculated as:

$$\frac{10}{2(\text{no of criteria} - 1) + 1}\left(X - \text{no of criteria} + 2\right)$$

This must be rounded to the nearest integer to give the actual weighting, W.

This formula is derived from a generalised formula for moving the axis of a normal distribution.

In the example shown in Figure 4.2 therefore the weighting of the criteria for selecting a book are:

Price	3
Length	4
Subject	9
Author	7

Step (iv): Awarding Scores to Each Criterion

The criteria established so far are those by which we wish to evaluate the hardware required. The next step is to decide on what factors the hardware is to be judged for each criterion. These factors must be defined and assigned scores; the range 1 to 10 is usually satisfactory, where 10 is the best score and 1 the least favourable. For instance, in the book example I may decide that the factors on which to judge the subject matter are Thriller, Spy Adventure, Science Fiction, other.

These I award the following scores:

SF	10
Thriller	7
Spy Adventure	7
Other	2

This implies that I really want a science fiction story, but don't really mind a thriller or spy adventure and will only buy another type of book as a last resort.

There are two types of scoring scheme that can be used, namely a sliding scale and a fixed scale. In the example above a fixed scale is used, fixed scales can also be defined where each score represents a range of values. Sliding scales are best where it is difficult to quantity the criterion. As all scoring schemes should be devised before the models which are to be evaluated are selected it is possible that no model will score 10; indeed they may all be weak in this criterion and score low marks.

Step (v): Evaluate Each Model

Each manufacturer's model to be evaluated can now be analysed to see how well it satisfies each criterion. Steps (i) to (iv) must be complete before this stage is embarked upon in order to make sure that no bias slips into the process.

For each criterion consider which factor the model best agrees with and award it that score. When each criterion has been dealt with, multiply each score by the weighting for that criterion and total these results. This gives the overall score for that model. For instance, if one of the books I was considering was a James Bond spy thriller I would award it 7 under the criteria 'Subject'. I would then multiply this by the weighting for 'Subject' which is 9, to give me 63. I would then add this to the numbers obtained for the other criteria to give me the final score for that book.

Step (vi): Comparing Results

Once Step (v) has been carried out for each model under consideration, the model with the highest score wins. If several models have high scores, which are close, then the final selection must be made as objectively as possible. There is no rigorous method which can be applied but various factors can be taken into account which help make the final decision. These are discussed in Chapter 10.

A summary of an example of the whole process is given (table 4.1) for the selection of a portable radio, a task which most people will have undertaken at some stage in their lives. The complete worked example is given in Appendix 5. Chapter 11 is a case study of the hardware selection exercise where the whole process is worked through for the various items of data communications hardware discussed.

The Overall Selection Process

In selecting data communications hardware it is not sufficient to select the best model of each individual item; the entire system must be considered as a whole before the final selection is made. Several factors must be taken into account; namely, compatibility, ease of installation, maintenance procedures and cost. This overall selection process is discussed in detail, in Chapter 10.

GENERAL CRITERIA

Delivery Time

In the initial planning process a date by which the data communications system has to be installed and operating will have been decided. Therefore allowing a reasonable time for installation and testing, it will be possible to fix the required delivery date for the hardware. What is a reasonable time? This depends very much on the complexity of the system and on the number of suppliers involved: no hard and fast rules can be given.

In general, hardware suppliers will quote delivery times of about 13 weeks, but the Post Office can take up to a year to supply the necessary circuits. Delivery can either be to one central site with the client taking the equipment to each installation site or the supplier will deliver to all installation sites as necessary. It is important to identify which is to occur so that sufficient time can be allowed in the delivery schedules.

It is frequently necessary to deliver all the equipment to a central site initially, for testing. This also allows a pool of equipment to be formed so that some systems testing can be carried out before the communications links are available. If the network is to be installed in phases then phased delivery of equipment may be appropriate.

It is often wise to select a required delivery date and insist on it by making delivery date an essential criterion. Alternatively if there really is flexibility in the implementation plan then meeting the date could be desirable but not essential. In that case a scoring system could be used such as:

delivery within timescale	10
delivery 1 week after	8
delivery 2 weeks after	6
delivery 4 weeks after	4

delivery 8 weeks after	2
delivery more than 8 weeks late	1

What level of complexity is selected depends on the individual situation. However by setting a realistic delivery date it should be possible to use a simple scoring system using two or three levels. If it seems necessary to use more levels than this, look at the delivery date again and re-evaluate its accuracy.

Simplicity of Installation

Factors to be considered under simplicity of installation are:

- type of power supply required;
- weight of the hardware and consequently any restrictions on where it can be located if floor strengthening is required;
- any special facilities required, eg air conditioning, false flooring;
- location of telephones or PO cables for modems;
- any restrictions on length of cabling to adjacent hardware; for instance, modems or terminals to a multiplexer;
- whether the supplier takes responsibility for the installation;
- security;
- proximity of equipment which might cause interference.

Any of these factors may prove to be an essential criteria under varying circumstances. As desirable criteria they should be scored in such a way that the simpler the installation the higher the score.

Reliability

Modern electronic equipment is basically reliable and is often designed in a modular fashion so that if a failure occurs part of the equipment continues to function. Reliability is extremely important in data communications because generally the system is closely integrated into the work of the company: when a failure occurs it is immediately felt by staff who cannot perform their jobs. It is usually prohibitively expensive to provide complete redundancy of hardware so equipment must have a high level of reliability.

The level of reliability of various items of hardware varies with its role in the network. For example if a front-end processor should fail then the entire network is out of action, but if a remote modem for a teletype fails then only the terminal will be unable to access the host. Such factors will influence the importance a user will attach to this criterion.

Even if reliability is not inherently important in the data communications system it is far more satisfactory and far less work to have a network that functions happily without frequent failures. Most hardware manufacturers quote the reliability of their equipment in terms of Mean Time Between Failure (MTBF). It is important to discover from the supplier figures that have actually been achieved by other users of the hardware in a situation similar to the one in which it is to be used, rather than theoretical figures or figures achieved under ideal testing conditions.

It is difficult to suggest an acceptable MTBF for different items of hardware because of the diversity of equipment on the market. Therefore possibly the best scoring system that can be assigned here is one in which a sliding scale is used in relation to the MTBFs, the higher the MTBF the higher the score.

There may be situations where a long MTBF is essential in which case an essential criterion of an MTBF greater than X hours can be defined and all equipment which fails to meet this standard rejected. A high MTBF over this limit could still be judged as a desirable criterion, with the highest score going to the hardware with the longest MTBF.

Maintenance Support

There are two aspects to this problem:

a) *The level of maintenance support from the supplier of the hardware.* Before evaluating this criterion the user must decide how important the data communications system is and how much money should be spent to ensure high availability. The user may decide that the system is so important that duplicated hardware must be provided at all locations so that if a failure occurs service is lost only for a very short time. In this case the speed with which the supplier responds is not vital. Similarly if the system, or a part of it, can be out of action for a few days without causing problems the speed with which the supplier responds is of little significance. Most users however fall between these two extremes.

   Generally the maintenance contracts offered by suppliers cost more the shorter the promised repair time and the better the service. Most major suppliers offer a range of maintenance contracts. One major company, for example, offers six maintenance plans, details of which are given in Table 4.1.

It is generally true that the major suppliers with many locations throughout the UK can turn out within four hours, while smaller companies cannot offer this level of service in all parts of the country. Some suppliers do not provide a maintenance service and require the customer to identify the faulty equipment and return it for repair.

The importance of the level of maintenance support will therefore vary from one selection process to another. The criteria relating to it should be set up in such a way that they reflect the value of maintenance to your particular organisation. If for example a link must not be out-of-action for more than 8 hours then a maintenance contract that can achieve this becomes an essential criterion, while the turn-out time and cost of the maintenance contracts within this limit become desirable criteria (as it will still be necessary to select the best value for money from those companies which can meet the essential criterion). In deciding on the turn-out time required, remember to allow additional time for the repair to be effected. For example if a link must not be out-of-action for much more than 8 hours then it is necessary for the maintenance staff to arrive within 5–6 hours.

If for instance an organisation decides that a link cannot be out-of-action for more than 12 hours, but preferably is not out-of-action for more than 8 hours, then a suitable scheme would be:

Essential criterion:

Maintenance staff arrive and repair in less than 12 hours

Desirable criterion with a scoring scheme similar to:

Maintenance staff arrive and repair in 8 hrs or less	10
Maintenance staff arrive and repair in 8–12 hrs	5

b) *The ease of access, built-in diagnostic features and ease of replacement of components,* ie factors which make fault-finding and maintenance procedures easy for both user and supplier. No matter what an organisation's maintenance requirements, ease of maintenance procedures and built-in diagnostic facilities are an advantage, especially for an organisation with little experience in data communications. If however a piece of hardware offering these features is considerably more expensive than the equivalent without then the matter must be considered closely. It may well prove more economical to buy two pieces of hardware with few frills and keep one as a standby than to buy one with a wide range of diagnostic features backed up by an expensive maintenance contract.

An opinion on this criteria can be given on a 10 point scale where:

10 = Procedures are quick and simple

1 = Maintenance and repair involve lengthy and difficult operation

Post Office Regulations

The written consent of the Post Office is necessary before any apparatus not supplied by the Post Office is used in association with a Post Office installation. Details of Post Office requirements for the various categories of privately owned and maintained equipment normally encountered are contained in Publications known as Technical Guides. A list of the subjects issued by the guides and the address of the Post Office Telecommunications Headquarters Services Department from which copies may be obtained are given in Appendix 3. Care must be taken in this area as Post Office approval relates to the use of the equipment and not to the equipment itself. For example a modem may be approved for use on a 4800 bps private circuit but further approval would be necessary for its use on the PSTN.

The purpose of such restrictions include prevention of interference with other users of the carriers' facilities, and prevention of damage to installation or injury to maintenance staff.

The Post Office will evaluate for a fee a proposed attachment which does not already have approval, but this process can be extremely lengthy. The information to be submitted with each application varies according to the service with which the item is to be associated. Most manufacturers do not make equipment that does not have Post Office approval generally available unless it has been used in other countries where it has approval. It is nevertheless important that the question of Post Office approval is raised with the supplier and a satisfactory answer received. The Post

Office itself keeps a list of all equipment currently approved so a check can be made from this if necessary.

Having Post Office approval is probably one of the most important criterion for any item of data communications hardware. It is therefore always an essential criterion.

Compatibility and Standards Requirements

There is an obvious need to ensure that all the equipment in the network is compatible. This can be achieved by buying equipment that is directly compatible or by buying some sort of bridging device which enables equipment with different characteristics to work together. The latter path is not recommended, especially to inexperienced data communications users, as all sorts of unforeseen problems can arise.

International and manufacturers' standards exist for communications hardware and it is important to ensure that equipment conforms to the most relevant of these. The most important international standards in the present context are the V series CCITT recommendations described in Chapter 2. By ensuring that a modem, for instance, conforms to Recommendation V24 and V28 the systems designers can be sure that their terminals will be compatible with the computer at a hardware level. International standards however are sometimes not as standard as is thought. This is because most such standards offer options and a user must check that not only does the device adhere to the standard but what options in this standard have been selected.

Manufacturers' standards can prove more of a problem. These standards, at both hardware and software levels, can be restrictive in that choice is limited to those pieces of equipment that are compatible or that can emulate compatible equipment.

Compatibility with other equipment in the network is an essential criterion and it is recommended that directly compatible hardware is given preference over use of some sort of bridging device.

Supplier Factors

There are four aspects to be evaluated in this area. These are:

- is the supplier credible and financially stable?
- is the product range going to be developed? (If not, there could well be problems when additional or replacement items are required in the future.)

– is the hardware of proven design?

– do you already have any equipment from this supplier?

Any of these could be essential criteria in some circumstances but more usually one would expect them to be no more than desirable. The first three aspects cannot be analysed in any quantitive way. It will be necessary to assign scores in the range 1–10, to each of these criteria (where 10 is the most favourable), in as objective a fashion as possible. For instance a supplier who had been in business for many years with a good reputation would score 10 for the first criterion, while a supplier who had been in business only six months and was still building up his business might score only 1 or 2. Both might achieve a high score for the third criterion however.

It is advantageous to keep the number of suppliers to a minimum. The fourth criterion will therefore require a straightforward yes or no, with the positive reply scoring 10 and the negative 1.

Costs

Costs must be evaluated in several areas, namely:

– basic cost of the hardware;

– installation costs;

– maintenance costs, per annum, for the level of maintenance you have decided is necessary;

– any other costs (these must be specified before awarding scores);

– can equipment be bought, leased or rented?

Each of these criteria must be scored separately. The basic cost should include the cost all essential features while 'other costs' could include charges for features which are desirable but may appear as optional on some model; the cost of achieving compatibility with other equipment or any other extra changes which might be incurred. Any of these four desirable criteria could have upper bounds on the amount of money that an organisation is prepared to spend. In such case the upper bound becomes an essential criterion.

The method by which the item can be paid for, ie the last criterion, can often be a deciding factor in the selection process. This is especially true where capital expenditure is carefully controlled, making purchase difficult, but where equipment rental can be paid from overheads.

SPECIFIC CRITERIA

These criteria relate uniquely to each type of data communications hardware. They are connected with the specific purpose of the equipment and the manner in which it carries out this purpose. As with the general criteria, these will be essential or desirable or of no interest whatsoever depending on the circumstances in which the equipment is to be used. In the following chapters a list of the specific criteria for each hardware type is given. Each criterion is discussed and scoring schemes are suggested.

SUMMARY

Table 4.2 summarises the steps to be taken in the evaluation method. Figure 4.3 is a complete list of general criteria which must be considered for each item of hardware.

Table 4.2 Summary of steps in evaluation exercise

	Step
1	Draw up list of criteria on which the evaluation is to be based. General criteria Specific criteria
2	Identify essential criteria
3	Assign weightings to remaining criteria
4	Award scores to each criterion
5	Evaluate each model
6	Compare results and make selection.

Figure 4.3 Complete list of general criteria

*	Delivery time
*	Simplicity of installation
*	Reliability, mean time between failure
*	Level of maintenance support
*	Ease of maintenance procedures
***	Post Office approval
***	Compatibility
*	Credibility of supplier
*	Product range to be developed
*	Hardware of proven design
*	Any equipment already from supplier
*	Costs

*** = Essential criteria always

5 Modems

INTRODUCTION

A modem is required at both ends of each line in a data communications network and as a result is the most common piece of hardware in any network. It is therefore not surprising that the successful operation of such a network depends very much on the performance of its modems. The manner of the modem selection must reflect this importance; it is not an easy task because of the enormous range of equipment available and its complexity.

In Britain the Post Office places certain restrictions on the choice of modems open to users. Any modem must have Post Office approval for connection to Post Office circuits. In the case of the PSTN, modems are supplied by the Post Office itself; approval for the use of proprietary modems can only be obtained when the Post Office itself cannot supply a suitable modem. On leased lines the Post Office grants approval for the use of proprietary modems which meet appropriate specification and technical guidelines laid down and does itself supply a range of modems within its Datel services. These restrictions can significantly reduce the problem of modem selection.

In Chapter 2 the modem and its use were described briefly. This section expands on the definition given there. It describes the various modulation techniques, the different types of modems available, in what situation each of these is most suitable and the specific features of a modem which a user must consider when carrying out an evaluation exercise. It is unlikely that in any one network all the modems will be the same. The user must consider the network configuration and decide what sort of modem is required on each link. In the extreme case the evaluation exercise will have to be repeated for each pair of modems in the network.

MODULATION TECHNIQUES

A modem is designed to convert the digital signals of computer equipment into analogue signals for transmission along a communications link designed for voice traffic, and vice versa. There are several techniques available to carry out this modulation/demodulation process, the three main methods being amplitude, frequency and phase modulation. These are described briefly, below. Further details can be found in the *Handbook of Data Communications* (NCC Publications).

– Amplitude Modulation (AM)

In this modulation method, digital data is converted into an analogue signal by modifying the amplitude of a basic 'carrier' waveform according to the data being transmitted. The carrier is usually a signal of constant frequency suitable for transmission over the telephone system. As digital information consists of only two states, 0 and 1, two amplitudes are required, for example the '1' being set at a higher amplitude than '0', as shown in Figure 5.1. This method of modulation is sometimes called Amplitude Shift Keying (ASK).

– Frequency Modulation (FM)

In this method, two alternative frequencies are used to represent 0 and 1, the frequency of the signal being switched from one value to the other as data changes (fig. 5.2). This is sometimes known as Frequency Shift Keying (FSK).

– Phase Modulation (PM)

Each point in a continuous, repetitive signal can be identified uniquely by the 'distance' from a fixed reference point in the waveform (usually a peak or zero value). This 'distance' is known as the phase of the signal at the point under consideration. By analogy with a sinusoidal waveform, phase is measured in degrees where one full cycle contains 360 degrees. Changes in phase of a carrier signal can be used to carry information in the same way as changes in frequency or amplitude. For example binary '1' can be indicated by a phase change of 180 degrees and binary '0' by no phase change (fig. 5.3). This technique is also known as Phase Shift Keying (PSK).

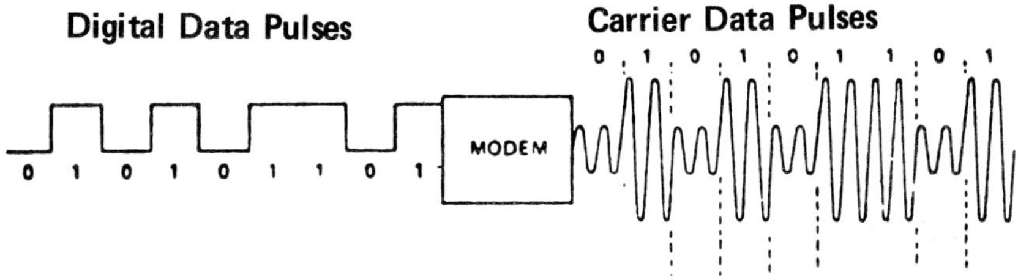

Figure 5.1 Amplitude modulation

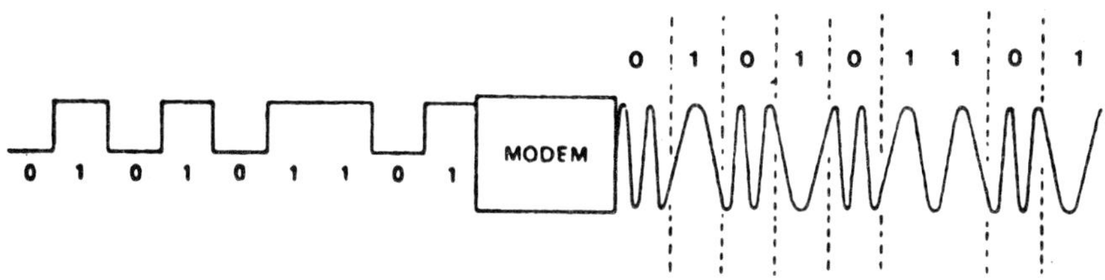

Figure 5.2 Frequency modulation

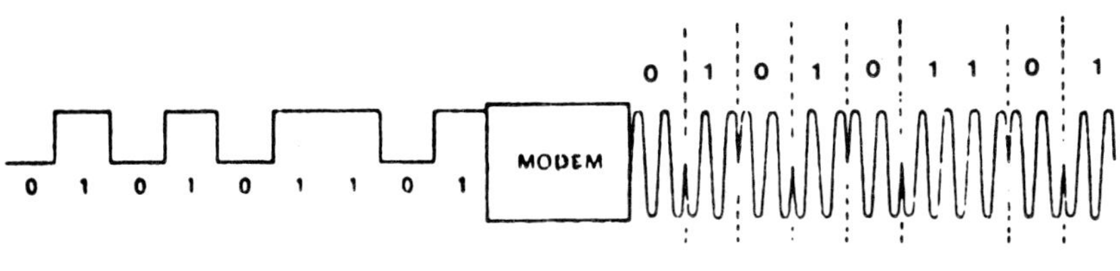

Figure 5.3 Phase modulation

– Multi-level modulation techniques

Transmission speeds higher than 2400 bps are achieved by making each signal element carry more than one item of information. To do this, multi-level modulation techniques must be employed which use combinations of the three main modulation techniques. Two ('dibits'), three ('tribits') or four bits of data are 'packed' into one signal element to produce the higher transmission rates of 2400 bps and above, the number of bits packed into one signal element depending on the bit signalling rate. For example, data transmission at 9600 bps is generally achieved by using a 2400 baud signal and packing four bits into one signal element. The technique used here is called 'quadrature amplitude modulation' (QAM) and is a combination of both phase and amplitude modulation. It is with these higher data transmission rates, therefore, that baud and bits per second (bps) values do not necessarily agree.

Amplitude modulation is the cheapest method of modulation but is less resistant to noise than the other methods. This technique is not widely used in modern modems. Phase modulation has better resistance to noise than amplitude and frequency modulation but it is a more elaborate technique. There are CCITT standards for the modulation techniques to be used at various speeds. These are listed in Table 5.1.

ACOUSTIC COUPLERS AND BASEBAND MODEMS

In addition to conventional modems there are two special classes of equipment used to perform a similar function. These are the acoustic coupler and the baseband modem.

Acoustic Couplers

'Acoustic coupling' is a method of data communication on the public network without using a conventional modem. This is a flexible method of establishing a data connection at the lower data rates and is widely used at 300 bps and below. The acoustic coupler has a special 'cradle' into which fits a telephone handset. The data signals are converted into tones which are picked up by the telephone microphone (this is the acoustic coupling) and transmitted on the line. (In the case of a modem electrical signals are transmitted directly.) Similarly signals received from the line are reproduced as tones from the telephone handset. The acoustic coupler converts these tones into digital data signals for use by the

Recommendation	Speed	Type of circuit
V21	200 baud	PSTN
V23	600/1200 baud	PSTN
V26	2400 bps	4-wire leased circuit
V26 bis	2400/1200 bps	PSTN
V27	4800 bps	Leased circuit
V27 bis	4800	Leased circuit
V27 ter	4800/2400 bps	PSTN
V29	9600 bps	Leased circuit

Table 5.1 CCITT recommendations for modulation techniques at various speeds

Modem type	Cost (purchase)
200 bps	£ 300
2400 bps	£ 500 – £ 600
4800 bps	£1600 – £2000
9600 bps	£3000 – £4000
Acoustic Coupler (300 bps)	£ 150 – £ 200

Some manufacturers offer discounts if more than one pair of modems is bought.

Delivery, installation and maintenance charges are usually extra.

Table 5.2 Modems costs, examples

terminal. Since no modem is used and since there is no permanent hard-wire connection, the equipment is portable and can be used anywhere there is a public network telephone handset. The error rate can however be higher than with modem connection.

Generally an acoustic coupler is only used at the terminal end of a line, a conventional modem being used at the other end. In order to work in this fashion however both devices must be compatible, and adhere to the same standards.

Although there is no electrical connection, acoustic couplers could still interfere with PSTN signalling and for this reason some PTTs prohibit their use on the PSTN. The Post Office permits, but does not encourage, their use in Britain, and does not supply them. They may be obtained from terminal and modem manufacturers.

Baseband Modem

The baseband modem does not perform a conventional modulation/demodulation function but modifies the digital signals for transmission over a physical cable pair. As they do not contain complex circuitry for modulation or equalisation they are less expensive than conventional modems.

As DC, baseband, signals are used, these devices require wide bandwidths and are therefore used for short, (for example up to 25 Kms) private in-house lines. It is not usually possible to lease them from PTTs and very few are approved for use on PTT supplied circuits. In Britain, the Post Office does supply a baseband modem, the Modem 8, which is used in conjunction with the Modem 9 on wideband circuits as part of the Datel 48K service.

IMPORTANT QUESTIONS IN MODEM SELECTION

Before establishing the criteria necessary to select the modems needed in a network, several important questions on the nature of the network must be answered. These are discussed below:

What Computer System and Application is the Modem Required For?

Before selecting modems some thought must have been given to transmission speeds, based on the nature of the application which will use the network. For example if the application requires large volumes of data to be processed over a short period of time then it is likely a high transmission speed could be required. This process is to a certain extent iterative as it is unusual for only one speed to be suitable. For instance the most suitable speed for the application may be in the range 2400–4800 bps and the final decision on which speed to use could be taken because the preferred modem operates at 2400 bps.

What Terminals Are to be Used?

The characteristics of the terminals should already have been identified by this stage. The modem must be compatible with the terminal so the operating features of the terminal must have been clearly defined. Most important in this area is whether the terminal operates in asynchronous or synchronous mode and whether the protocol is duplex or half-duplex; the latter has an effect on the importance of turnaround time.

Are the Links to be PSTN or Private?

Some links in the network may use the PSTN and other private leased lines. This will have an effect on the characteristics required of the modems, particularly in the area of Post Office approval. If a leased line is to be used with the PSTN as fallback two modems may be necessary. For only one to be used, it must not only have the ability to operate the fallback arrangement but Post Office approval for use on both leased and PSTN lines.

Are Modems to be Post Office, Proprietary or a Mixture?

If the PSTN is to be used and the Post Office can supply suitable modems then the user has no choice but to use them. This may well result in no further decisions on modems having to be made. However in many circumstances there may be several factors still to be resolved. For example, if the transmission speed required is 2400 bps over the PSTN the user has the choice of two Post Office services:

- Datel 2400 dial up using Modem 7C;
- Datel 2412 using Modem 12A.

What Types of Modems are Necessary?

It may be necessary for acoustic couplers or baseband modems to be used in the network. Acoustic couplers are particularly useful for low-speed teletypewriter use on the PSTN as they are cheap, portable and fairly reliable. For such a simple device no lengthy selection process is necessary, most terminal manufacturers supplying basically similar equipment. Baseband modems, on the other hand, could well prove useful on any short in-house links that may be necessary. For instance in a factory complex it may be required to link terminals to the computer, which is also on the site, but 2 or 3 kilometers away. In such a case the user can install his own metallic pair of wires and use baseband modems thus achieving significant cost savings. On Post Office circuits however the use of baseband modems is generally limited to cases where high transmission rates are required over very high bandwidth circuits. Like acoustic couplers, baseband modems are fairly simple devices and selection should be straightforward – preferably from a manufacturer supplying other devices for the network.

SPECIFIC CRITERIA FOR MODEM SELECTION

Figure 5.4 lists the criteria which are important in modem selection. The user must decide for himself which are important in his particular circumstances, although some, such as error rate will be important to most organisations. In general the modems available today do not vary enormously within one speed category and unless some special feature is required, it is likely that some of the general criteria, such as cost, delivery date or maintenance support, will be the deciding factors in selection. Where a scoring scheme is not suggested for a criterion, a straightforward scheme whereby the best model under evaluation is awarded 10 points and other lesser scores is generally suitable.

Table 5.2 lists typical model costs

Transmission Speeds

There are four widely recognised categories of modem transmission speed:

– low speed	–	up to 1200 bps
– medium speed	–	1200 bps 2400 bps
– high speed	–	4800 bps 9600 bps
– very high speed	–	48,000 bps 50,000 bps 64,000 bps 72,000 bps

Figure 5.4 Criteria for modem selection

*** Speed
*** Equalisation
*** Synchronous/asynchronous operation
* Diagnostic facilities
* Fallback features
* Voice adaptor facility
* Autoanswer
* Turnaround time
*** Error rate
* Operation, size, weight and colour
* Multiport feature
* Mounting
* Delivery time
* Simplicity of installation
* Reliability (MTBF)
* Maintenance support
*** Post Office regulations
* Ease of maintenance procedures
* Credibility of supplier
* Product range to be developed
* Hardware of proven design
* Any equipment already from the supplier
* Costs
*** Compatibility and standards requirements

*** = Essential criterion always

Most modems can operate at a number of transmission speeds, selection being made by a switch on the modem itself. Some modems can even be modified on site to operate at a higher speed than the highest possible when the modem was acquired.

The speed at which the modem can operate must match that required by the application. It is frequently useful however to have a modem which can operate at a range of speeds. Firstly, if for some reason an increase in line speed is considered in the future it will not be necessary to get a new modem, and, secondly, if the modem can operate at a lower speed this can be useful for data transmission to take place on a degraded line, eg PSTN fallback. This is generally only possible within categories of modem speeds. It is an essential criterion that the modem operates at the necessary speed in any selection exercise. However a range of speeds could also be a desirable criterion, for example:

Modem operates at necessary speed and at an alternative speed	10
Modem operates at necessary speed	7

Equalisation

On any line there exists a variety of characteristics which have an adverse effect on data transmission. There may be various noises and forms of distortion, (eg phase jitter), attenuation (loss of signal) and delay. To compensate for these inconsistencies in transmission, lines can be conditioned or 'equalised' by the Post Office to make them more suitable for transmitting data. The process of equalisation deliberately introduces opposite characteristics to those inherent in the line, which results in the line characteristics being made 'flatter' (fig. 5.5). However, this cannot compensate for noise.

Equalisation can also be carried out by the modem itself. This is especially necessary for data rates of 2400 bps and above, where line equalisation alone is not sufficient; the higher the data transmission rate the better the quality of line required. PSTN lines are not usually of the same quality as private circuits; they are not conditioned in the way that private circuits may be. It is therefore, important that for the higher data transmission rates some degree of equalisation be carried out by the modem itself.

Modem equalisation can be carried out in several ways.

– 'Compromise' equalisation

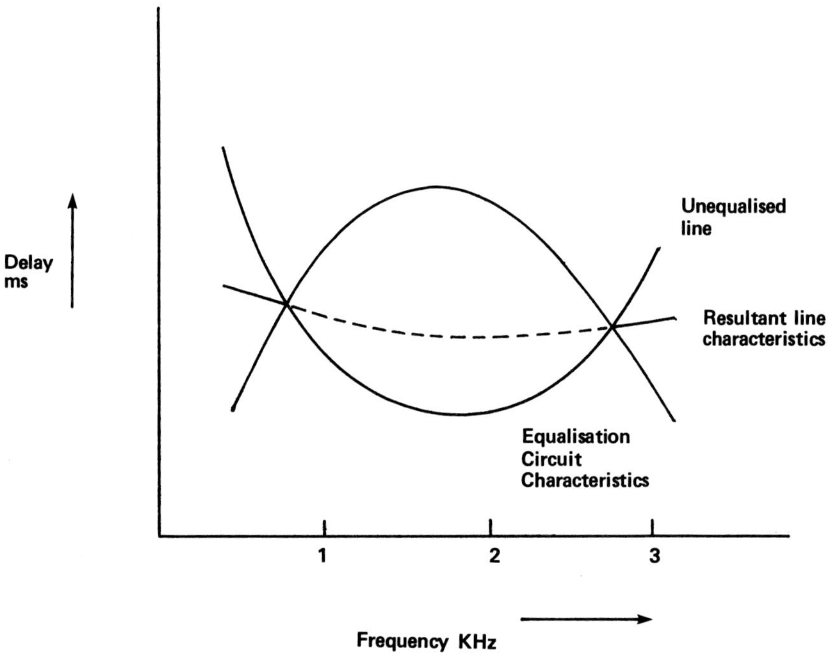

Figure 5.5 Effect of equalisation

An equaliser circuit which has fixed characteristics to provide a satisfactory standard of transmission is incorporated into the modem. 2400 bps modems often have facilities for several alternative equaliser circuits depending upon the line conditions experienced when the modem is installed. However, this method of equalisation is not usually sufficiently accurate for speeds above 2400 bps.

– 'Variable' equalisation

For speeds above 2400 bps variable equalisation is used instead of 'compromise' equalisation. Variable equalisers are adjusted on-line to individually compensate for the actual line conditions encountered.

There are two types of variable equalisers:

– Manual equalisers

The modem has a number of controls which are individually adjusted when the modem is installed to adjust the optimum delay and attenuation characteristics across the frequency band. This adjustment provides a composite circuit within the modem which exhibits the opposite characteristics to those of the line.

This method is now not used very widely as it means checking and possibly re-setting the equaliser whenever the line characteristics change which could be every two or three days.

– Automatic adaptive equalisers

No operator attention is required. Equalisation is achieved by a signal processing circuit which automatically adjusts itself to the line characteristics. The modem thus continuously adapts itself to any line changes which occur during transmission. This type of equaliser is more convenient and is indeed essential in situations where the modem is unattended or on a site where no suitable operator is available to carry out manual equalisation.

The scoring scheme here depends on the speed of the modem. At 2400 bps or less, a suggested scheme is:

Equalisation circuits in modem	10
No equalisation circuits in modem	1

At 2400 bps or above a suggested scheme is:

Automatic equalisation in modem	10
Manual equalisation in modem	6
Compromise equalisation or no equalisation	1

Synchronous/Asynchronous Operation

The modem must match the synchronous or asynchronous operation of the terminal it serves. This is an essential criterion. Synchronous modems are more complex than asynchronous ones because of the additional circuitry needed to achieve synchronous transmission. If synchronous operation is to be used, there are several alternatives open to the user in choosing from where the synchronising signal or clock is obtained:

- the terminal can provide it;
- the computer can provide it;
- one or other or both modems can provide it.

It is usual for the modem at the central site to provide the clock but there are many exceptions. Hence many synchronous modems have facilities for selecting the method of clocking, eg Internal or External clock, Master or Slave clock.

Asynchronous modems on the other hand can serve synchronous terminals by making use of the synchronising clock in the terminal.

Diagnostic Facilities

It is important to be able to find a fault quickly and to discover whether it is the terminal, modem, circuit, or any other piece of hardware or software that is causing the breakdown. Therefore some kind of diagnostic equipment, although not an obligatory feature of a network, is certainly extremely useful, particularly in the more complex networks.

Diagnostic facilities can be built-in to the various pieces of network hardware, either centrally and/or at the remote sites. Many modems are themselves equipped with diagnostic facilities. These include:

- Loopback mode

  Loopback mode allows the user to test both the modem and the link. Some loopback facilities enable the remote modem to be checked also.

The transmit lines of the circuit are connected back on to the receive lines so that the modem receives the same data that it is transmitting. A comparison can then be made between the received data and the transmitted data and the performance of the modem then determined.

This can be done at both the data interface between the terminal and the modem – digital loopback – and the line interface between the modem and the line – analogue loopback.

– Other self-testing facilities

The modem will usually have other self-testing functions besides the loopback facility, eg signal quality and line level display, and bit error detection which test the quality of the line and the number of bits transmitted in error respectively.

Such indicators can frequently be extremely useful but are not essential.

The user must decide for himself how much he is prepared to pay for diagnostic features in a modem. The higher the speed of the modem and the network itself the greater the worth of such features.

Fallback Function (Standby)

A modem used on a private circuit can be equipped with a 'fallback' facility so that in the event of the failure of the private circuit, the modem can be switched to operate on the PSTN. This is only possible if the modem is Post Office supplied or if it has Post Office approval and can operate on the PSTN at speeds above 2400 bps. (Below this speed, Post Office policy regulations dictate that a Post Office modem is required.) The modem may have to operate at a reduced speed on the PSTN (eg 1200 bps instead of 2400 bps) if the line quality is not good enough to support operation at the original speed. For this reason the modem must be capable of operation at a range of speeds. If a particular feature is required of a modem which can only be supplied by a proprietary modem manufacturer, and a PSTN fallback feature is also required, it may be necessary for the user to rent a second, Post Office, modem at the lower speed to supply the fallback transmission via the PSTN.

By this stage the user should have planned the fallback requirements and will know at what points in the network, modems with a fallback function are necessary. The scoring scheme then depends on whether the feature is required – there is no point in paying for an expensive extra feature if it is not necessary.

Voice Adaptor Facility

The voice adaptor enables a private circuit to be used for voice transmission instead of data transmission. Data transmission is interrupted by disconnecting the modem from the circuit and connecting a proprietary telephone handset to the line instead. A switch on the modem effects this change.

Voice transmission over the link is very useful, if not essential; it allows engineers or operators to speak over the line to set up transmission or to check its integrity during any fault finding sessions. If the user decides this feature is desirable then the only possible scoring scheme is:

Voice adaptor facility	10
No voice adaptor facility	1

N.B. it is possible for the P.O. to provide suitable signalling facilities and a P.O. telephone handset on a leased circuit as an alternative to a voice adaptor.

Autoanswer Mode

Modems operating on the Public Network may incorporate an automatic answering facility. There are several ways in which this may be implemented. Basically the terminal equipment, ie computer or terminal, and its associated modem detect the ringing current of an incoming call on the PSTN line. The modem is then connected to the line and sends an answering signal that is received by the originating equipment. This simple 'handshaking' procedure then allows data transmission to take place.

This facility is particularly useful in the computer bureau environment as terminal users can dial into a computer port and be automatically connected to the computer without operator intervention.

CCITT Recommendation V25 details the standards for using automatic calling and/or answering equipment on the Public Network. However it is only necessary to use the V25 procedures when operating internationally. Many Datel modems supplied by the Post Office incorporate a simplified method of autoanswer within the modem, which is satisfactory for operation within Britain.

If the user decides this feature is desirable then the only possible scoring scheme is:

Autoanswer facility	10
No autoanswer facility	1

'Turnaround' Time

With half-duplex operation of a transmission line, the modems have to be switched from transmit mode to receive mode at the end of each block of information. This reversal process or 'turn-around' as it is known, requires a finite time, which can be considerable with some high speed modems. Turnaround time, sometimes known as 'Ready for Sending' delay, is generally taken to mean the time which elapses between the terminal passing a 'Request to Send' signal to the modem and the modem replying with a 'Ready for Sending' (or 'Clear to Send') signal, to indicate that it is able to accept data for transmission. This delay is caused because the two modems, once they have switched from transmit to receive, have to resynchronise and possibly re-equalise. The delay can vary from a few milliseconds to several seconds. Consequently some networks make use of full-duplex channels, even though the transmission protocol only requires half-duplex channels.

It can be seen that turnaround time is wasted as data cannot be transmitted during that time. Long turnaround times mean less throughput efficiency. If two modems are comparable in all features except turnaround time then greater efficiency will be achieved by selecting the one with the shorter turnaround time.

Error Rate

Error rate is the term used to indicate the number of errors that can be expected when transmitting a given block of data. It depends on both the quality of the line and the efficiency of the modem. An inferior modulation method for a particular modem speed for example may increase the error rate to an unacceptable level, as may using an unconditioned line for high speeds. By applying the latest technology, techniques for modulation and equalisation can be improved and the entire performance of the modem improved resulting in, amongst other things, lower error rates.

Block size is generally fixed by the characteristics of the computer. The block error rate is of more significance then the bit error rate in transmission. If a block is in error it will be retransmitted. A high block error rate could therefore result in a situation where no data is received as every time a block is transmitted it contains an error and must be retransmitted.

The Post Office will not guarantee an error rate on the PSTN although it may give some estimate of error rates for lower speeds. The telephone network in the UK is commonly taken to have an error rate of 1 error in 10^4, the range being from 1 error in 10^3 up to 1 error in 10^6.

In terms of modem selection, error rate is suprisingly not the most important criterion. It is difficult to establish the error rate of any particular modem without conducting precise and elaborate tests as it depends to such an extent on the conditions under which it is to be used.

Modem Design

Besides an awareness of the various modem techniques and features described previously, aspects of modem design also requires consideration.

- Modem switches, controls and indicators

  It is advantageous if these are easy to understand and operate.

- Size, weight and colour

  These aspects may be important to the user. For example, space may be very limited or a particular design or colour may be required to fit in with the computer room colour scheme.

Ease of operation is an important criterion in modem selection. Scores must however be awarded subjectively depending on the user's opinion of the device. If there are any special requirements in terms of size, weight or colour, then the obvious scoring scheme is score 10 if criterion is satisfied and 1 otherwise.

Multi-Port/Split-Stream Modems

Some 4800 and 9600 bps modems have an option which allows the multiplexing of up to 4 independent data channels, ie the modem takes on the role of a multiplexer and transfers several separate data streams at different speeds simultaneously over the one link. For example there may be one 7200 bps channel and one 2400 bps channel or alternatively four 2400 bps channels on a 9600 bps link.

The multiplexing option costs little compared to the cost of a multiplexer and it is becoming quite widely used. If this feature is useful or it is thought it might prove useful in the future then the suggested scoring scheme is:

Several options for splitting channels by switch selection	10
Several options for splitting channels by internal strap selection	7
One fixed option for splitting channels	5
No facility to split channels	1

Mounting: Case or Rack

Modems can usually be installed in either of the following ways:

- by rack mounting, ie housed together with a number of other modems or other data communications equipment, often a number of units sharing a common power supply.
- by case mounting, ie a single modem is installed in its own case.

Rack mounting is generally preferable at a site where several items of hardware are used, especially at the computer centre where there will be one modem for each link to the computer. This is because it is tidier and saves space. This feature will have to be decided on a modem-by-modem basis; most modems however may be mounted in either fashion.

GENERAL CRITERIA

The general criteria discussed in Chapter 4, namely:

Delivery Time
Simplicity of Installation
Maintenance Support
Post Office Regulations
Compatibility/Standard requirements
Supplier factors
Cost
Reliability

must be taken into consideration in the selection process as well as the specific ones discussed in the preceding section. Two important points must be noted with respect to modems however.

These are *Post Office regulations* and *compatibility/standards requirements.*

Post Office Regulations

Check carefully with the modem manufacturer on the status of Post Office approval. The Post Office may well have approved the modem for a particular configuration only, eg separate approvals are required for a 4800 bps modem to operate on the PSTN and on a private network.

The Post Office at the time of writing is in the process of developing a 4800 bps service over the PSTN (Datel 4824). Currently 4800 bps modems are not provided by the Post Office who will, therefore grant a licence to use approved proprietary 4800 bps modems on the Public Network.

Compatibility/Standards Requirements

International standards exist for modem operation, as well as interface and transmission standards, in the form of the CCITT 'V' series of recommendations (see Appendix 4). Despite this apparent standardisation, to which many manufacturers conform, there is still a variety of techniques in use, particularly at the 4800 bps and 9600 bps operation levels. For example, the modem manufacturer may use his own modulation technique. It may be superior to the methods defined by the CCITT and employed by the Post Office, but it may make for compatibility difficulties if different modems are used on the network. This is particularly important in the case of international networks.

There are two CCITT standards which special attention must be paid: V24 and V28 (equivalent to US standard EIA No. RS232). This is the standard which lists the interchange circuit requirements between data terminal equipment (DTE), ie the terminal, and the data circuit terminating equipment (DCE), ie the modem. It is the only standard to be accepted internationally by all manufacturers.

A connector consisting of 25 pins, is used at the interface between the modem and terminal. The electrical characteristics of all the interchanging circuits made through this connector conform to CCITT standard V28. However, a modem will not make use of all the circuits (usually between 8 and 18 of them) and so the user must check which particular circuits of the 25 available are being used, and to which pins they are connected in the particular modems to be rented or purchased.

6 Multiplexers and Concentrators

INTRODUCTION

Multiplexers and concentrators both serve the same purpose in the data communications network: they increase the efficiency of the network's performance and thus reduce data communications costs. The variety of multiplexers that can be obtained is wide, ranging from a simple hard-wired device costing several hundred pounds to a sophisticated intelligent multiplexer costing several thousand pounds.

Concentrators tend to be even bigger, more powerful and more expensive. Because of the similarity in their use it is appropriate to discuss multiplexers and concentrators together in one chapter. In some circumstances the user may be faced with the initial decision – do I need a multiplexer or a concentrator? before thinking of selecting a particular model.

MULTIPLEXERS

A multiplexer takes a number of communication channels and combines the signals into one common channel of a greater bandwidth or data rate in such a way that the original signals can be extracted again. A multiplexer is required at the end of each line to multiplex and de-multiplex the signals (fig. 6.1).

All multiplexers, except some of the intelligent ones, provide a 'transparent' connection in the network; they do not interfere with the data in any way. Therefore the existence of a multiplexer in the network does not necessitate the modification of any software.

Types of Multiplexer

Simple Multiplexer

This device is an unintelligent unit which performs the basic role

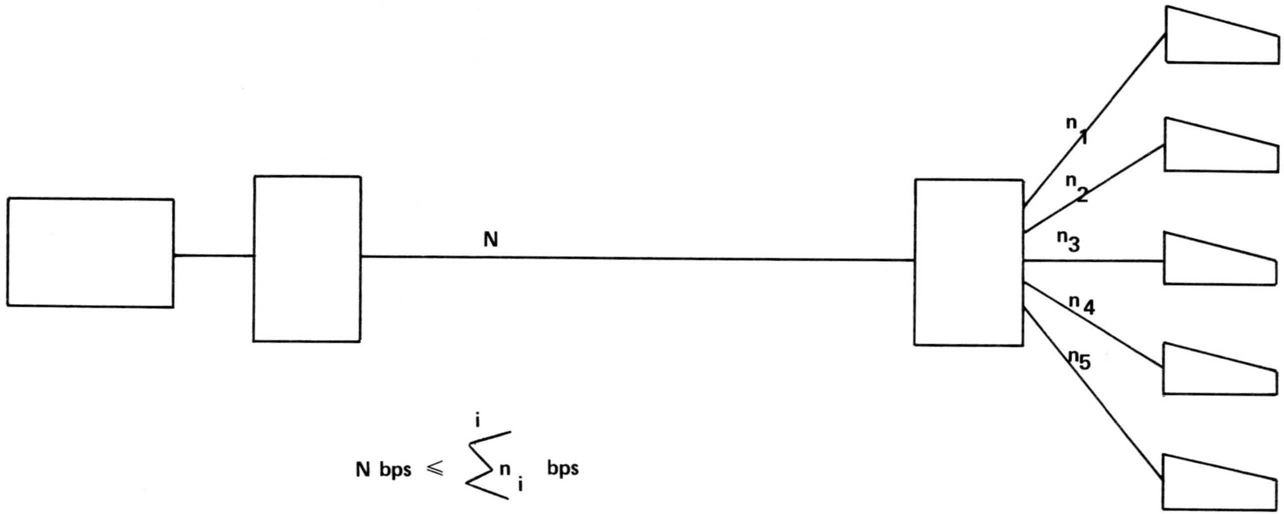

Figure 6.1 Use of a multiplexer

of reducing total communications line costs and has no other function.

It uses a single line to carry data from a number of terminals, presenting the data from the terminal to the computer as if each terminal has its own direct wire connection to the host. This is obviously cheaper than each terminal having its own line connected to the computer.

There are two types of simple multiplexer: frequency division multiplexers (FDM) and time division multiplexers (TDM).

Intelligent or 'Software' Multiplexer

This type is equipped with intelligence, ie it is programmable. It is usually a microprocessor or minicomputer, and besides carrying out the functions of the simple multiplexer, may also have some of the features of a concentrator.

The difference between this kind of multiplexer and the concentrator (see below) is not always clear. The intelligent or 'software' multiplexer is a hybrid device, combining both multiplexer and concentrator features.

The intelligent multiplexer is sometimes called a 'dynamic' or 'statistical' multiplexer, and is based on time division multiplexing.

Multiplexing Techniques

There are two methods by which a simple multiplexer simultaneously transmits data from a number of terminals along a single line to the computer.

Frequency Division Multiplexing (FDM)

This technique divides the line's bandwidth into a number of individual channels, one for each terminal. Each of these channels is assigned a portion of the total frequency bandwidth (fig. 6.2). For example, a 2-wire circuit with a bandwidth of 3000 Hz could be divided into 12 channels each with a bandwidth of 240 Hz. No modems are required as the multiplexer itself carries out the modulation/demodulation process.

FDM is the oldest multiplexing technique. It has been used for many years by PTTs in dividing physical pairs of wires into a number of separate voice channels for telephone communication. It was the first multiplexing technique used for data traffic.

The channels themselves are separated across the frequency

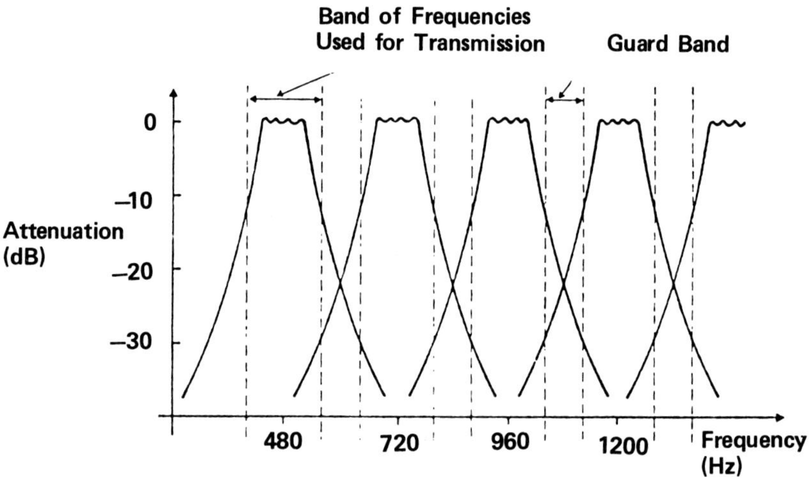

Source: Handbook of Data Communications

Figure 6.2 Frequency bands of a Frequency Division Multiplexing system

band by 'guard-bands' or safety zones to prevent cross talk or overlapping of the signals. This means that the available bandwidth is under-utilised and wastage is therefore incurred. A further limitation of this method is that channel capacity is restricted by the bandwidth of a voice grade telephone line (3400 Hz) so that this method can only be used for a limited number of low-speed channels. For example, for speeds of 50 bps, the lowest speed used for data transmission purposes, it is possible to have 24 individual channels. This is the maximum number of channels possible at this speed because 60 Hz bandwidth is needed for each channel for transmitting data with 60 Hz each side of the channels to act as guard-bands.

Data transmission at higher speeds can be allocated to the voice grade bandwidth in the same way but the number of channels will decrease as will the number of 'guard bands'.

Various CCITT recommendations exist for assigning frequencies and guard-bandwidths in FDM. These form part of the R series of recommendations on telegraphy and include specifically:

R31 for systems employing amplitude modulations at 50 baud

R35 for systems employing frequency modulation at 50 baud

R37 for systems employing frequency modulation at 100 baud

R38A for systems employing frequency modulation at 200 baud

Frequency Division Multiplexing is not used as much as Time Division Multiplexing (see below). When FDM is used, it is often as a speech and data multiplexer (SDM), when the circuit is used simultaneously for transmitting speech and data. The upper part of the bandwidth is restricted for data transmission and the lower portion is used for transmitting speech (fig. 6.3).

Another use of FDM is as a means of constructing a multidrop configuration. An FDM multiplexer sensitive to only one frequency can be located at each site in the configuration with a multiplexer sensitive to a range of frequencies at the computer site (fig. 6.4).

Although relatively inefficient and inflexible compared to TDM, FDM is a cheap and simple multiplexing method, suitable for a limited number of low-speed channels.

Time Division Multiplexing (TDM)

This method allocates each terminal a period of time on the line. These individual time slots are collected together into what are known as 'frames' to form a single high-speed data stream for

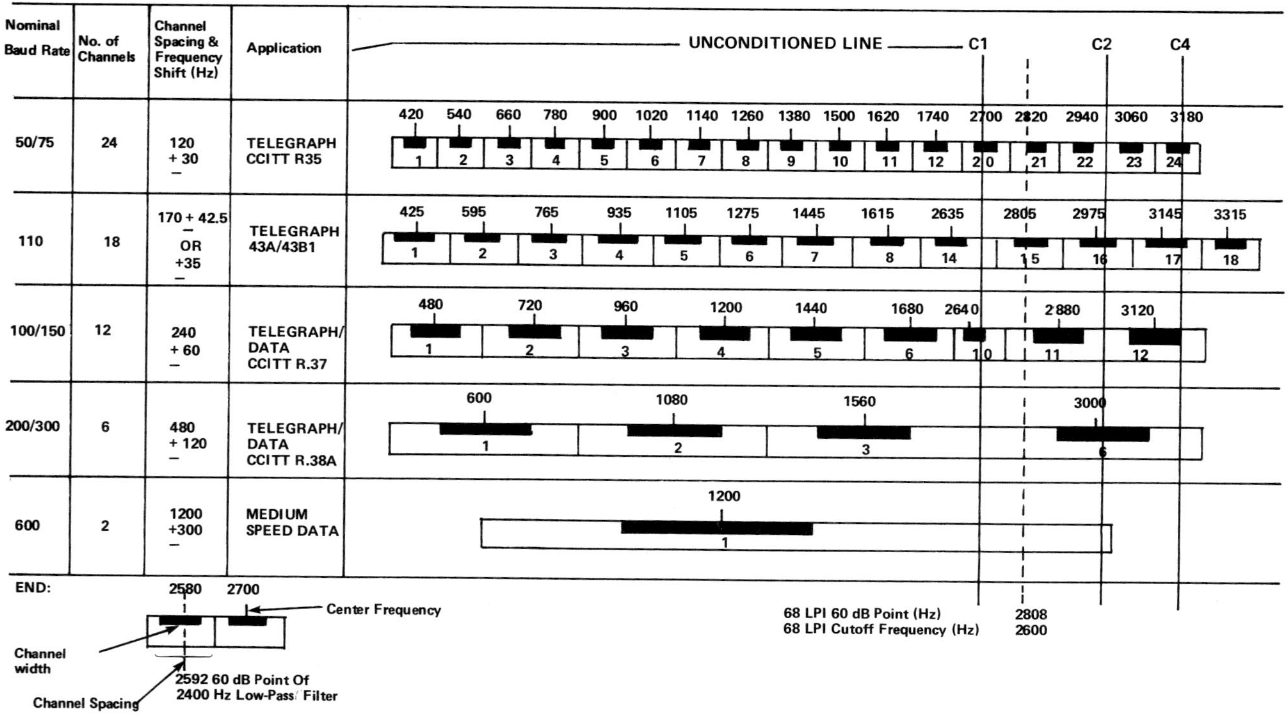

Figure 6.3 Standard frequency assignments

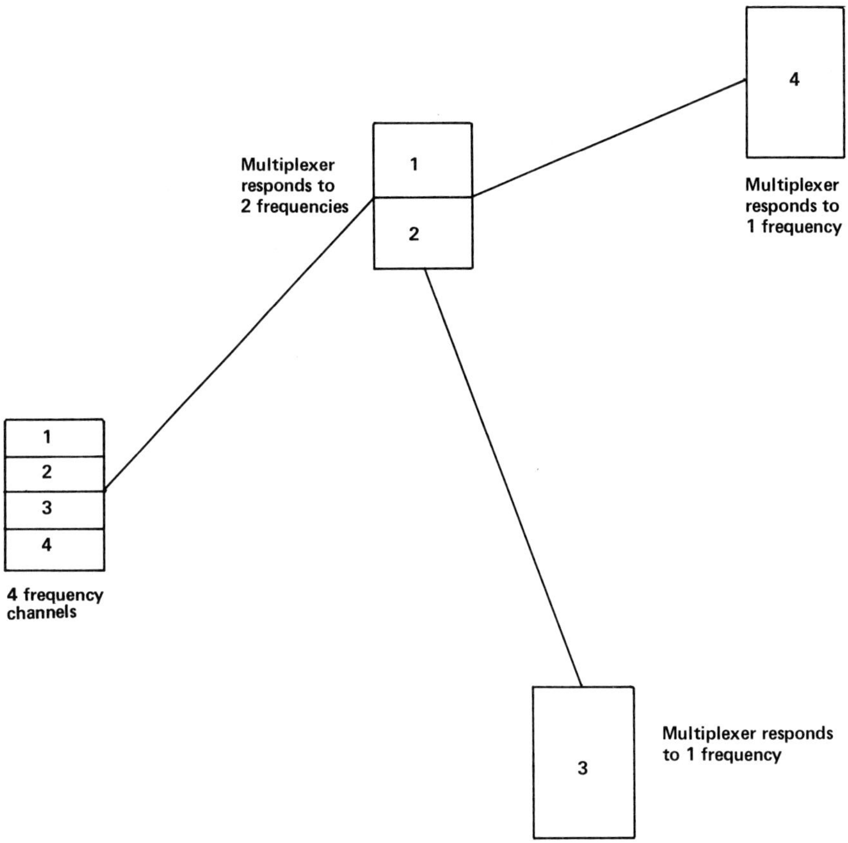

Figure 6.4 Multidrop configuration

transmission along the line (fig. 6.5). Each frame therefore contains data from each of the terminals attached to the multiplexer. Figure 6.6 shows a typical use of a TDM multiplexer for terminals of different speeds. High speed modems are required at each end of the link. The data can be transmitted along the multiplexed channel in one of two different ways:

– 'Bit' TDM

  The data extracted from the terminals' low speed lines leading to the multiplexer, is interleaved by time on the high speed line on a bit by bit basis (fig. 6.7). The slots are separated by a unique framing pattern or synchronisation character. Each slot a predetermined number of bits, 'idle' bits being transmitted if a terminal has no data to send. This way of manipulating the bits keeps the multiplexers at each end in synchronisation with each other.

– 'Character' TDM

  With this method, instead of bits being interleaved for transmission over the single link, a character made up of a predetermined number of bits is taken from each low speed terminal input channel to make up the 'frames', the frames then forming a single data stream for all the terminal channels (fig. 6.8).

  A complete character is received and assembled in a 'serialiser'. It is then stored in a buffer for transmission along the high speed line in the terminal's allocated time slot. As with bit TDM, if a terminal is inactive an idle character will be transmitted in its allotted time slot. Each 'frame' of time slots also contains a synchronising character, as well as frame control information which gives such details as message length and addressing. As whole characters are interleaved in each frame rather than bits, the frames are longer, but there is a lower proportion of synchronising characters.

– Bit TDM versus character TDM

  Bit TDM is used with terminals and modems operating synchronously as synchronous transmission is commonly bit oriented in its transmission method, whilst character TDM is used for terminals and modems operating asynchronously as asynchronous transmission is basically character oriented.

  With character TDM, as the characters are buffered before transmission it is possible to remove the start and stop bits

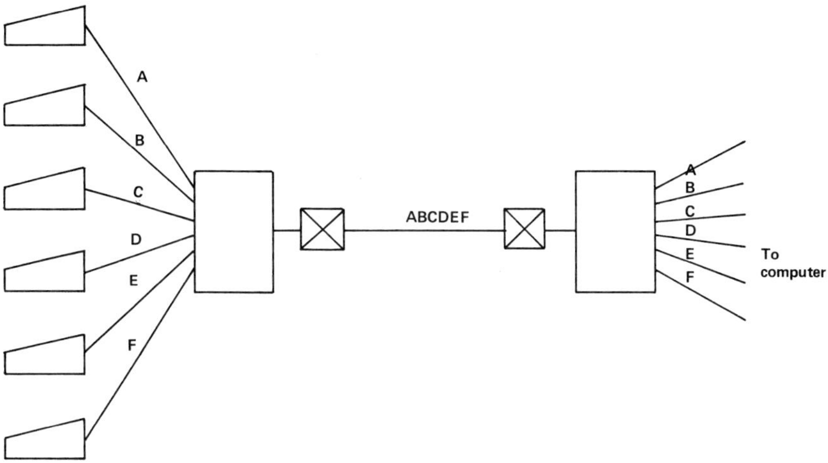

Figure 6.5 Time Division Multiplexing

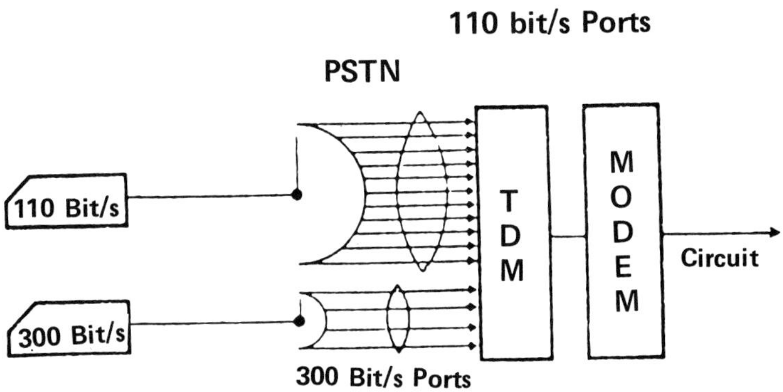

Figure 6.6 Access into a local TDM from different speed terminals, using the PSTN

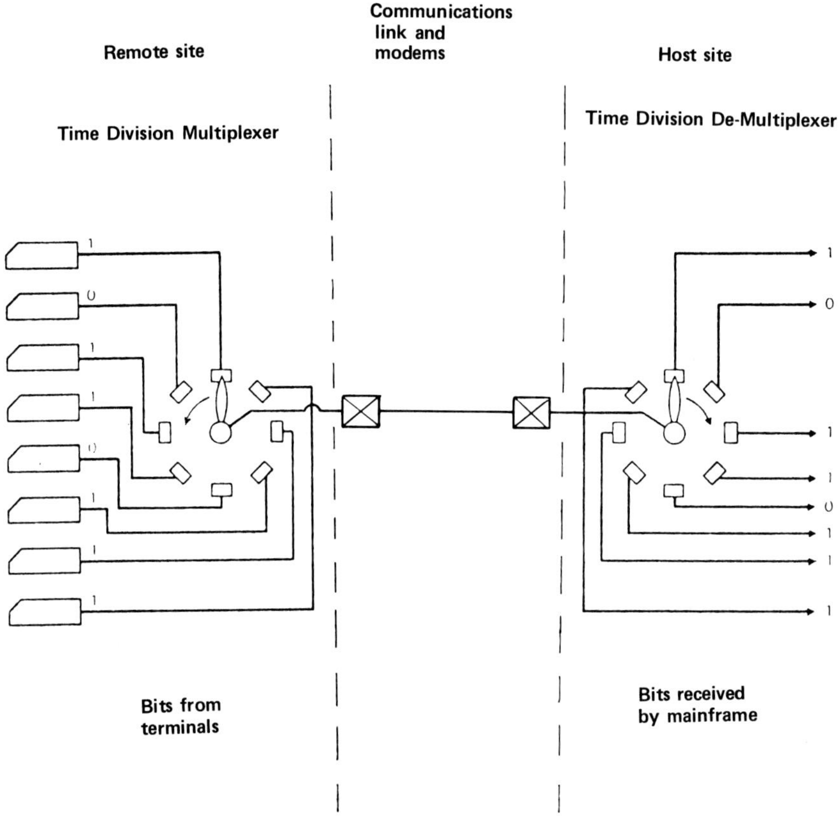

Figure 6.7 Principles of bit interleaved time division multiplexing using simple commutators

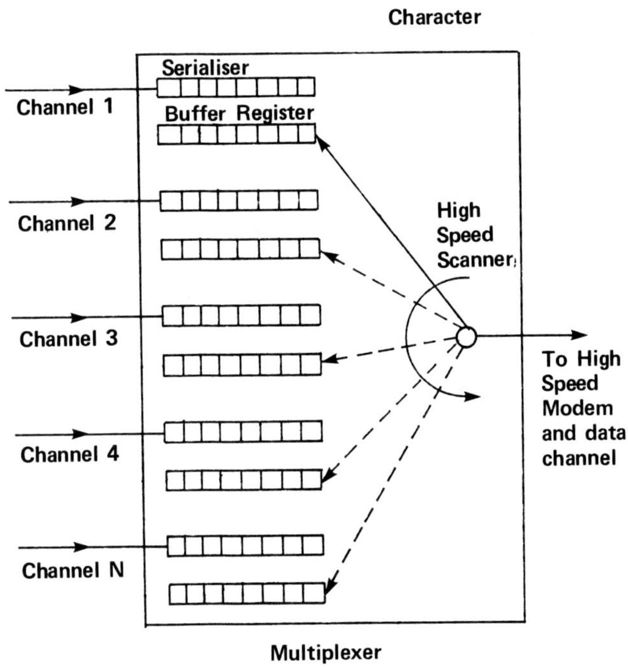

Figure 6.8 Simplified character interleaved time division multiplexing

('stripping') from each character over the high speed transmission link and for them to be reinserted by the multiplexer at the other end. This means that more data can be sent over the high speed link thus increasing the apparent efficiency of the multiplexer. The proportion of SYNC bits in character TDM is lower in character interleaving than bit TDM which also means that more data can be transmitted. On the other hand extra storage is required for character TDM as each character is buffered before transmission whilst bit TDM requires only the minimum of storage.

Character TDM has an advantage over bit TDM with respect to noise or error bursts. An error burst within a bit TDM frame could affect all the bits in the frame. As each bit emanates from a different terminal a number of terminals will be affected. If a character frame contains an error burst the frames either side could be affected thus affecting only three terminals.

Character TDM tends to be used for slower speeds, up to 1200 bps, whereas bit TDM is mainly used for higher speeds, from 1200 bps upwards because low speed transmission is character oriented.

'Statistical' Multiplexing or 'Dynamic Bandwidth Allocation'

A major weakness of the simple multiplexing techniques described above is under-utilisation of the line. With FDM, bands of frequencies are allocated to terminals whether the terminals are active or not: in the TDM method, time slots within frames are allocated to terminals which may or may not be busy. To achieve greater efficiency some method is required which will allocate time slots only when terminals are active and thereby achieve much higher line utilisation. Such a method is 'statistical' multiplexing which uses 'dynamic bandwidth allocation'. A terminal is only allocated a time-slot when it has some data to send; the time-slot allocation is carried out dynamically. This is the key feature of all intelligent multiplexers.

The method relies on the assumption that not all terminals will be active at any time. Using this technique means that the high-speed line bit rate is less than the aggregate bit rate of the terminals' low speed channels, usually between 20% and 50% less. This makes for greater throughput efficiency and increases the number of terminals which can be served. More buffer storage is needed to cope with any system overload at peak demand time and extra logic is necessary for controlling and addressing the data. If the system is overloaded, usually data from low-priority terminals is stored or rejected until the overload situation has passed. This process will be scarcely noticeable to the user.

The method implies some degree of contention at these peak times as the terminals are all competing with each other in their transmission of data.

Multiplexer Features

Although multiplexers are usually bought 'off-the-shelf' from the manufacturer and are not specially made for a particular system, they tend to be constructed in a modular fashion. A multiplexer usually consists of a basic unit with certain standard features such as a power supply while optional features may be added as required. They can be added on to the basic unit or chassis by slotting in additional electronic circuitry on boards; these are generally called 'cards'. This modularity therefore allows considerable flexibility in multiplexer selection.

Generally this will result in a list of essential criteria being drawn up with the remaining specific criteria being optional features which are desirable. These can then be scored on a simple system of 10 points if the feature is available and 1 point if it is not.

Frequency Division Multiplexers

Figure 6.9 lists the criteria upon which an FDM multiplexer should be judged.

- Input handling capacity

  The number of terminals, or inputs that can be supported is limited by the bandwidth of the voice grade line and the need for 'guard-bands' and depends on the speeds at which the terminals to be used operate. The maximum number of inputs that can be supported is 24, operating at a speed of 50 bps. The higher the speed the lower the number of terminals that can be supported. It is always essential that the multiplexer can satisfy the number of outputs required.

- Terminal input/output speeds

  Any speed up to 1200 bps is possible and different speeds can be mixed. It is essential that the multiplexer can deal with the input and output speeds required.

Figure 6.9 Criteria for the selection of a frequency division multiplexer

*** Input handling capacity
*** Terminal input/output speeds
* Speech and data multiplexing
* Monitoring and test facilities
* Mounting
* Design
*** Good filters
* Delivery time
* Simplicity of installation
* Reliability (MBTF)
* Level of maintenance support
* Ease of maintenance
*** Post Office regulations
*** Compatibility
* Credibility of supplier
* Product range to be developed
* Hardware of proven design
* Any equipment already from the supplier
* Costs

*** = Essential criterion always

– Speech and data multiplexing

With simultaneous transmission of speech and data, the upper frequency bands are reserved for data and the lower frequency bands for speech. This feature is particularly useful where an organisation has private telephone voice circuits linking different parts of the organisation. These circuits can then be adapted for simultaneous data transmission by using an FDM multiplexer, without incurring extra line rentals. The savings, particularly in the case of international circuits, can be substantial.

– Filters

FDM works by using filters to divide up the frequency spectrum. The design of the filters is critical to the number of channels that can be filtered into a given bandwidth and the maximum speed obtainable within a channel. With well designed filters the guard-bands can be smaller and there is less chance of crosstalk or co-channel interference. Good filters, ie filters which have a sharp cut-off at the desired frequency and allow frequencies below to pass without attenuation, are therefore an essential criterion of an FDM multiplexer.

– Monitoring and testing facilities

The frequency division multiplexer does have such facilities but they are very limited compared to those offered by the time division multiplexer.

– Mounting

A common option is for the multiplexer to be housed in a stand-alone 'table top' case or to be fitted for mounting into a rack.

Time Division Multiplexers

Figure 6.10 lists the criteria on which to judge a TDM multiplexer.

– Input handling capacity

TDMs can handle varying numbers of input channels or terminals. This maximum depends mainly on the speed of each terminal channel and the speed of the high-speed bearer circuit.

Multiplexer brochures will often indicate the various combinations of input channels rates possible, usually giving a chart with the details (fig. 6.11 for a typical example).

As with a frequency division multiplexer a suitable input handling capacity is an essential criterion.

– Terminal input/output speeds

Terminals operating at different speeds are usually acceptable. Generally, it is necessary for the multiplexer to know in advance the speeds that it will handle and it is also usual for groups of channels and their associated computer ports to be dedicated to a particular speed.

Figure 6.10 Criteria for selecting a Time Division Multiplexer

*** Input handling capacity

*** Terminal input/output speeds and adaptive speed control

*** Terminal input/output codes

* Monitoring and test facilities

* Mixed asynchronous/synchronous channels

* Mounting

* Design

* Delivery time

Simplicity of installation

* Maintenance support

* Ease of maintenance

* Reliability

*** Post Office regulations

*** Compatibility

* Supplier credibility

* Product range to be developed

* Proven design

* Any equipment already from supplier

* Costs

*** = Essential criterion always

(a) a small FDM

Combination	A	B	C	D	E	F	G	H
Port 1	50	75	110	300	300	300	300	600
Port 2	50	75	110	50	75	110	300	–
Port 3	50	75	110	50	75	110	–	–
Port 4	50	75	110	50	–	–	–	–

(b) a larger TDM

High speed links at
2400 bps

Number of inputs	Speed of inputs in bps
59	50
28	110
21	134.5
7	300

High speed links at
4800 bps

Number of inputs	Speed of inputs in bps
70	50
58	110
44	134.4
15	300

High speed links at
9600 bps

Number of inputs	Speed of inputs in bps
70	110
70	134.5
35	300
17	600

Number of inputs in table above are calculated using data plus non-data bits. Higher rates are possible if non-data bits are excluded from the main data channel. For calculation of numbers of channels, see "Transmission Channel Capacity" section of this leaflet.

Figure 6.11 Typical input handling capacity of multiplexers

With some TDMs, certain channels can accept any speed within a predetermined range. This is known as 'adaptive speed control', 'automatic bit rate' (ABR) or 'autobaud'. ABR is used chiefly in computer bureaux operations where a variety of terminal types may compete for access to the bureau computer. This can be a useful optional feature.

– Terminal input/output codes

Many simple multiplexing systems can accept data from terminals using different codes, eg EBCDIC, ASCII, CCITT Alphabet No 2. As with input/output code it is usually necessary for the codes to be specified in advance. Certain multiplexers have some or all channels which can accept any code within a predetermined range.

Again this facility is particularly useful for computer bureaux.

– Monitoring and testing facilities

Checks and tests can be carried out on the multiplexer and the individual multiplexer channels. These facilities can also aid error diagnosis of associated equipment such as modems.

Error counting can be carried out on the low-speed terminal lines; loopback switches can be provided to demultiplex multiplexed data for testing purposes; there can be lamps to indicate line and control signal activity and alarms to show certain conditions such as out-of-synchronisation, modem carrier failure and buffer overflow.

These features can be extremely useful but do add to the total cost of the multiplexer.

– Mixed synchronous and asynchronous channels

Some multiplexers are equipped with both character multiplexing and bit multiplexing logic and are thus able to cope efficiently with a mixture of synchronous and asynchronous lines.

– Mounting

As with an FDM multiplexer a TDM multiplexer may be designed for mounting in a rack or fitted in a separate case for table top use.

– Multiplexer design

It is an advantage if the multiplexer has clearly marked controls which are easy to operate and positive in their response.

- Multi-node configuration

  Instead of supporting only one high-speed line as happens with simple multiplexers, a number of high-speed lines can be supported. This enables an intelligent multiplexer to be used as a network 'node'. This can provide economic and flexible operation, allowing interconnection between many parts of the network. The great advantage of this is that an alternate route is then generally available in the event of line failure.

- Code/speed/protocol conversion

  Some networks may require code, speed or protocol conversion. For example, a number of terminals may be linked to two different host computers and be only compatible with one of them in data code and protocol type. Conversion to the required code or protocol can be carried out on selected channels to ensure compatibility. Multiplexers which have this conversion option are non transparent devices because they are manipulating or changing the data in some way. Special care must therefore be taken with network design.

- Error correction and detection

  Using an automatic repeat request (ARQ) system error free data is transmitted, the multiplexer sending the data retransmitting it if it is found to contain errors on first transmission. Thus error protection is provided over the multiplexer link.

- Port selection by terminals

  This feature enables data from a terminal to be routed to a particular computer port at the request of the terminal.

- Intelligent multiplexer at one end only (software demultiplexing)

Intelligent Multiplexer

Besides having the features of a simple multiplexer, the intelligent multiplexer has a number of other features which vary depending on its degree of intelligence. A complete list of features is given in Figure 6.12. The range of intelligent multiplexers is wide, some being more sophisticated in operation than others. The most intelligent of these are not always called 'intelligent multiplexers' by the manufacturer but are sometimes given names such as 'Intelligent Network Processor' or 'Intelligent Data Concentrators'.

Figures 6.12 Criteria for the selection of an intelligent multiplexer

*** Input handling capacity
*** Terminal input/output speeds and autospeed
*** Terminal input/output codes
* Network management
* Centralised network control
* Multinode configuration
* Code/speed/protocol conversion
* Error detection and correction
* Port selection by terminal
* Mounting
* Delivery time
* Simplicity of installation
* Maintenance support
* Ease of maintenance
* Reliability
*** Post Office regulations
*** Compatibility
* Supplier credibility
* Product range to be developed
* Proven design
* Any equipment already from the supplier
* Costs

*** = Essential criterion always

– Data compression

Data is compressed by transmitting the same information but by translating it into another code. This code usually has a variable bit length for different characters, with the aim of transmitting overall a smaller number of bits. This feature is particularly useful when dealing with frequently recurring bits or characters and can sometimes as much as double the data capacity of the high speed link. However the compression code is usually highly data-sensitive, and a code that gives good results for, say, a FORTRAN program listing may be highly inefficient with other types of data, eg numerical data. Thus the compression code must be carefully selected for the types of data being transmitted.

– Network management

Although the simple multiplexer performs certain monitoring and testing functions, the intelligent multiplexer can do much more. Statistical information can be provided to aid network management on such important factors as traffic density, retransmission rates, character error rates of terminals and line utilisation and line failure. Diagnostic features such as local and remote loopback are also provided. All this information can be used to detect any failure and to reconfigure the network as well as contributing towards efficiency in network operation.

– Centralized network control

An intelligent multiplexer can be provided with an operator console and a control printer. This feature can be used by a 'master' multiplexer for configuring the various nodes in the network avoiding the need for any operator intervention at the remote nodes. Network statistics can also be routed to the 'master' multiplexer for printing out using the control printer.

– Input handling capacity

It is possible for over 100 terminals to share one telephone line. Multiplexer brochures often provide charts which give examples of the multiplexer's capacity for handling a mix of terminals depending on their different speeds. The multiplexer's maximum capacity should be noted carefully as the multiplexer must be able to cater for possible growth of the user's network, but in terms of the number of terminals and maximum aggregate data rate.

– Terminal input/output speeds

The intelligent multiplexer can handle terminals operating at a number of different speeds. Usually the simple multiplexer has to know in advance which channels are to be allocated to what particular speeds. With most intelligent multiplexers some or all of the channels can accept any speed, ie employ autobaud or adaptive speed control. Some devices have an 'autospeed' option where the multiplexer can automatically determine the speed – and character format – and then dynamically configure the terminal ports for the rate determined.

– Terminal input/output codes

Intelligent multiplexers can accept a variety of data codes. As with input/output speeds, the code format is automatically detected and then routed to the appropriate port of the host computer.

In some cases it is possible to do without a multiplexer at the host computer end. There are two ways of effecting this: the remote multiplexer can have a custom-built module to emulate one of the protocols used by the computer; the host computer (or front-end processer) can be provided with multiplexing software.

This method is to be approached with great caution; it can be very expensive and complicated.

CONCENTRATORS

The concentrator collects data from a number of relatively low-speed channels and interleaves or 'concentrates' the data into bulk loads for transmission over a higher speed link to the host computer. A concentrator unlike a multiplexer always employs contention.

The concentrator also performs other functions besides this line economy activity. It is a programmable machine, usually a minicomputer, and can carry out various tasks such as network reconfiguration, message verification and error checking. The specific activities of the concentrator depends on how it has been programmed.

Concentrators are not usually purchased 'off-the-shelf'. They are generally obtained on a one-off basis, that is the hardware is purchased and special-purpose software is provided for the customer's own use. Often a supplier will do this on 'turnkey' basis, ie supply the whole system, hardware and special software, as a complete package.

Minicomputers are conventionally used as concentrators as they have the storage, power and intelligence for this purpose and are often designed specifically to work in a real time environment. Mainframe computers, sometimes used as front-end processors, would be too big and expensive to serve as concentrators.

As minicomputers requiring additional software concentrators are much more expensive than multiplexers and can cost up to several hundred thousand pounds.

Types of Concentrator

The word 'concentrator' can have many different meanings in communications but it generally refers to a message concentrator. It is this type of concentrator which will be dealt with in this section. There is also a device called a 'line concentrator' or 'line contention unit'. This is a simple hardwired device costing several hundred pounds and has none of the characteristics of the message concentrator. It is simply a line switching device which enables a greater number of input channels to share a smaller number of output channels on a 'first come, first served' basis.

Concentrator Techniques

(Message) concentrators employ certain techniques that can be called the basic characteristics of concentrators.

– Contention

  This technique is employed in all concentrators. A larger number of terminals or input channels contend or compete for a smaller number of computer ports or output channels. The bits-per-second rate of the high-speed line is less than the aggregate bits-per-second rate of the low-speed terminal channels leading to the computer. The terminals are all able to access the concentrator simultaneously, but they cannot all send data to the host at the same time, as happens with the multiplexer. They may have to queue depending on the available channel capacity in the high-speed circuit. The concentrator, therefore, takes advantage of the fact that terminals are not busy all the time.

– Polling

  Although the terminals contend for actual transmission of data to the host computer, competition is minimised by the fact that the concentrator carries out polling on behalf of the host. The concentrator asks each terminal in turn if it has any messages to send. The terminals can be polled in a particular sequence, some more frequently than others if necessary. By polling, the concentrator can actually control the terminals in their data transmission to the host and reduce queueing. As a contrast, multiplexers do not perform any polling functions.

– 'Store and Forward'/Queueing

As terminals all have access to the concentrator but cannot send data simultaneously a buffer is required in the concentrator to store the incoming messages to be routed to the computer when there is channel capacity available. The terminals therefore, may have to queue with their messages. The whole process is carried out in seconds, the queueing, and 'store and forward' activities are scarcely noticeable to the user.

– Non-transparency

Concentrators, in contrast to all multiplexers, except the very intelligent statistical multiplexers, are non-transparent. Data passing through the concentrator is manipulated by the concentrator and may have its format changed, although the content remains the same. This happens in the case of code compression or automatic speed or code conversion. Data is 'processed' for greater data communications efficiency. This activity by the concentrator affects other equipment in the network because, for the sake of compatibility they have to know that this activity is being carried out.

Concentrator Features

Concentrators have a number of common features but as they are devices which involve a high degree of customising, and the final product depends very much on the customer's requirements. This has a great effect on the method of selection. The best approach is probably for the required features to be identified by the year and then manufacturers invited to tender for a device with such features.

– Processing activity

As a non-transparent, custom-programmed device, the concentrator can carry out various processing activities which aim to transmit data from the terminals to the host computer in the most efficient, cost-effective way. In this context, therefore, processing refers to any changes made to messages by the concentrator, in order to improve the data communications activity. For example, the concentrator can carry out speed, code, and protocol conversion, error checking and message verification as well as data compression.

– Network management

The concentrator can also carry out certain network management activities, such as performance monitoring, line switching for system reconfiguration, gathering of information for statistical purposes and fault diagnosis.

MULTIPLEXER AND CONCENTRATOR SELECTION

The user, in reading the previous section, should now be aware of the similarities and differences between the multiplexer and the concentrator. There are basically four devices from which a choice can be made, ie a simple FDM or TDM multiplexer, an intelligent multiplexer or a concentrator. The user must first decide which of these is the most suitable before the manufacturers can be contacted. Table 6.1 summarises the main characteristics of each type of device, with indications of cost and the size network for which it is suitable.

Before making a choice between the different pieces of equipment it is necessary to answer the following questions:

- What are the system requirements? How is the network going to expand?
- What terminal types are being and will be used in the network?

It is also worth bearing in mind the prime functions of the multiplexer and the concentrator which are:

- to reduce network costs
- to improve data communications efficiency

The user's choice will be governed by the application requirements and by the network's present size and expected expansion. As a network develops, at some stage, multiplexing will often be required. As multiplexers are usually modular in construction, it is possible to start off with a multiplexer supporting say ten terminals and, as the network gets bigger, increase the number of channels that can be supported to say twenty, for example, by

	TDM	Bit TDM	Character TDM	Computer Concentrator
Efficiency — asynchronous channels	Good	Good	Very Good	Excellent
Efficiency— synchronous channels	N/A	Good	Good	Excellent
Channel capacity	Poor	Good	Good	Excellent
Flexibility	Poor	Good	Good	Very Good
Reliability	Good	Very Good	Very Good	Poor
Complexity	Good	Good	Good	Poor

Table 6.1 Costs and characteristics of multiplexers and concentrators

adding cards to the basic chassis. Multiplexer capacity, ie the maximum number of terminals and combinations of terminals and speeds that can be supported, is important so this must be carefully checked to cope with network expansion.

The various options available can be made use of, as the network gets bigger, by adding on modules to the basic multiplexer, eg adding a synchronous terminal to a basically asynchronous configuration. Again, not all multiplexers offer the same features so the options open to the user must be carefully checked.

It is only with a very large network that a concentrator should be considered. As previously stated they are very expensive devices. Which device is ultimately chosen depends very much on the network's specific requirements. There are no strict rules governing the choice between the different multiplexers or between an intelligent multiplexer and a concentrator.

7 Communications Controllers and Front-End Processors

INTRODUCTION

Some kind of interfacing device is required between the mainframe and the communications line (apart from the modem) in a data communications network. These devices vary enormously from a simple hardwired data adaptor unit, through the hardwired communications controller to a fully programmable front-end processor (FEP). Most manufacturers do in fact offer a range of interfacing devices for different types of data communications handling problems.

The first data communications systems relied on simple data, or line, adaptor units between the modem and the computer. One of these is required for each line, and the bulk of the communications handling is carried out by the central processor. This arrangement is perfectly adequate for a small number of terminals and is still used today in such cases, as well as for interfacing terminals to small machines designed for data communications such as minicomputers.

As the number of terminals increases however, there is a marked increase in central processor utilisation in order to support them. In the late 1960s mainframe manufacturers introduced hardwired communications controller units (CCUs) which removed some of the communications load from the mainframe. These were costly and inflexible however and today have been largely replaced by programmable devices, generally minicomputers, which perform the functions of the hardwired controllers and may additionally remove other facilities from the mainframe. Such equipment, the front-end processor as it has become known, is now available from all major computer manufacturers as a complement to their mainframe range.

The role that communications controllers and FEPs play in a communications network is discussed. They not only remove communications handling from a mainframe frequently ill-suited for the task, but can also give the user greater network flexibility and reliability. The equipment operates in a uniquely intimate relationship with the host computer, connected to it by a high-speed channel.

Consequently, the problems caused by using different suppliers are such that, quite understandably, only the more experienced users have been prepared to take the step. This chapter should enable the reader to assess the level of complexity required in the device which will interface the mainframe and the communications network, and facilitates product evaluation. Table 7.1 lists the products available from the major computer manufacturers.

LINE ADAPTOR UNITS

These terminate lines and, in the simplest case, perform serialising and deserialising of the bit strings as the line operates in bit serial mode and the channel in bit parallel mode. They may also be more sophisticated and perform a certain amount of hardware checking such as parity checks. A line adaptor unit can either be a stand-alone device which interfaces a link to a host mainframe channel or can be part of a more complex device such as a CCU or FEP. Each unit may support only one speed and one line protocol or signalling method. Frequently however it may be possible to set by software such parameters as line speed, number of start/stop bits, asynchronous/synchronous operations, etc.

HARDWIRED COMMUNICATIONS CONTROL UNITS (CCU)

To give good service to a terminal it will almost certainly need to be given a high priority in CPU resource allocation. This will mean that when a character arrives from a terminal other work requiring CPU time will be inhibited until this has been dealt with. The more terminals that are added, the more complex and numerous the tasks necessary to handle them and the less time the CPU spends doing other tasks. Hardwired communications control units, CCUs, appeared in the mid-sixties as the number of organisations using remote access terminals began to grow. These units are able to perform the simpler logical operations and reduce the load placed on the central processor by terminal handling.

The tasks performed by a CCU are character- and not message-based; typically, they include:

- parallel to serial and serial to parallel conversion of bits within characters (serialising and deserialising);

Manufacturer	Model
Burroughs	B866, B876 communications processors
CDC	255X series host communications processor Cyber 1000 data communications processing system
Honeywell	Datanet 6600 series integrated network processor
IBM	2701, 2702, 2703, 3704, 3705 Integrated communications adaptor
ICL	7903, 7904, 7905, 7906 Communications link controller, CLC Communications network processor, CNP Device control unit, DCU
NCR	621 Communications controller 721 Communications processor
Univac	Data communications processor (DCP) Communications Symbiont Processor (C/SP)

Table 7.1 Communications controllers and front-end processors available from the major computer manufacturers

– simple character and message error detection;
– identification of control characters;
– insertion and deletion of synchronising characters when synchronous communications is used;
– character framing with stop/start bits for asynchronous communications;
– providing the necessary interfaces with the communications lines via modems, or line adaptors and with the CPU input/output channel;
– automatic dialling and automatic answering of calls;
– automatic polling.

Certain optional features such as character code conversion are also possible.

Communications knowledge however still resides in the mainframe, which is aware of the physical configuration of the network. Therefore communications software, frequently called the Communications Control Program, CCP, is necessary in the host processor to supply facilities such as buffering and queueing, error control, code conversion and the collection and analysis of statistics related to terminals or line activity, errors and failures. Different computer systems implement a CCP in different ways.

DISADVANTAGES OF HARDWIRED COMMUNICATIONS CONTROLLERS

Most of the major computer manufacturers were marketing hardwired CCUs by the early seventies and there was also a large market in IBM plug-compatible devices. These plug-compatible models were frequently not completely hardwired but although they contained some software, modifications were not possible. It was the independent suppliers rather than the computer manufacturers themselves that led the move towards software devices which offered several distinct advantages over their hardwired predecessors.

The principal disadvantages of a hardwired CCU are:

– the high cost, typically in the region of £500 to £3000 per line, depending on configuration;
– because the CCU deals in characters and not messages the central processor still has to perform a large number of communications tasks; it is constantly being interrupted to

deal with full buffers or error conditions and may still have to cope with polling. As the data communications system grows, the pressure on CPU time frequently increases to the point where a more powerful host mainframe is required;

- communications control programs frequently take up large areas of store, leaving little room for other data processing work to be carried out;
- being hardwired, CCUs are necessarily inflexible devices and it is difficult for example to arrange support of new terminal types and new protocols. Data communications is still undergoing rapid change and hardwired devices are not ideal in a changing environment;
- if the host fails, a hardwired CCU cannot function or help to provide a graceful degradation of service to the terminals attached to it.

A software device, being more flexible, can help overcome some of these problems, and is generally cheaper on a cost/performance basis. Many hardwired devices are still in use however although few companies market them today.

FRONT-END PROCESSORS (FEP)

The programmable devices, sometimes having peripheral equipment of their own, are used to perform the communications functions for the main processor. Figure 7.1 shows the configuration of a typical FEP, which is generally a minicomputer designed for communications tasks with such features as efficient interrupt handling and good character/byte manipulation. Like all intelligent devices, an FEP can be programmed to do as much or as little of the work as required. Ideally an FEP would assume all terminal, line, buffering and message control functions allowing the main processor and the user application programs to treat the communications network as just another high speed on-line peripheral device. Frequently however they do little more than emulate a hardwired CCU. The latter type of FEP is known as a passive front-end processor and the former as an active front-end processor.

Passive Front-End Processor

Passive FEPs perform the same range of functions as the hardwired CCUs that they emulate. The reader may well, in that case, ask why they have generally replaced hardwired devices. There are two overriding reasons for this:

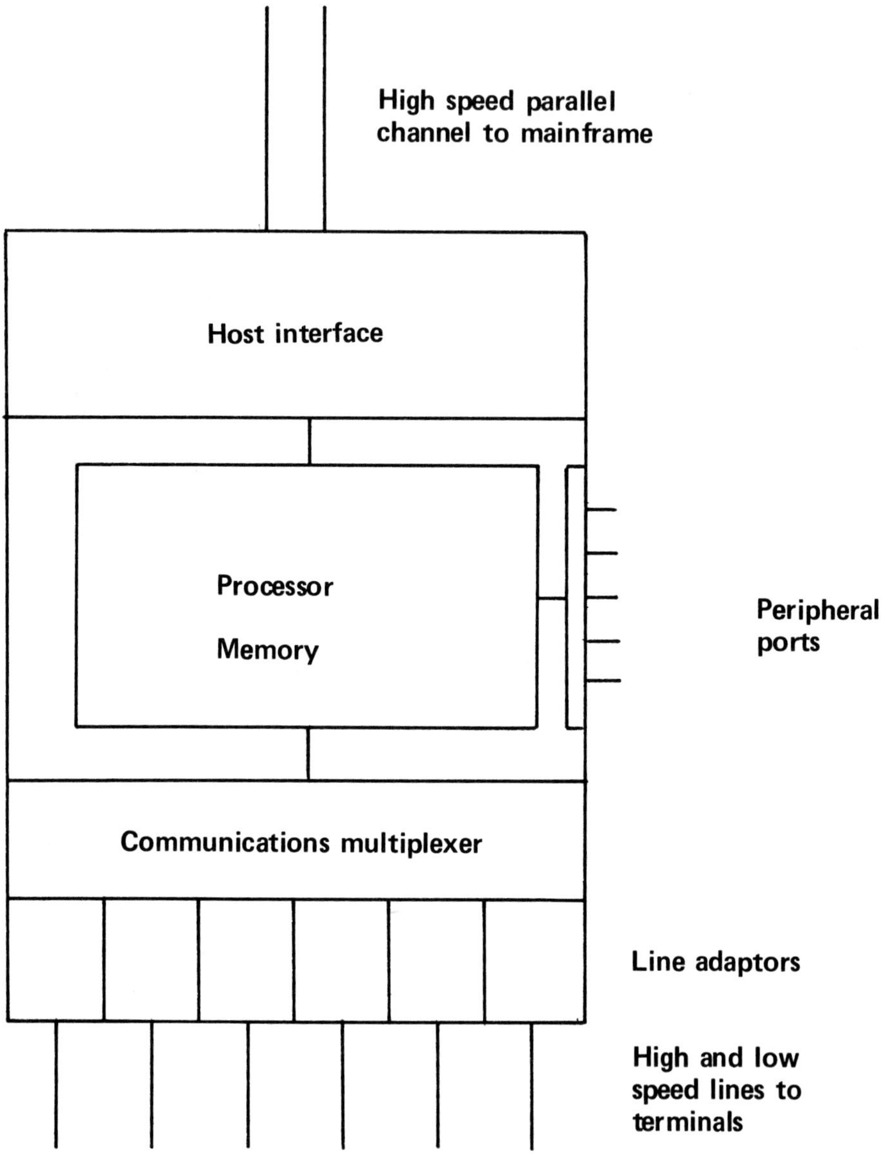

Figure 7.1 Typical front-end processor

- the continuing fall in on-line storage and processor costs and the rise in expertise in data communications software has made it cheaper to use a minicomputer than a hardwired CCU;
- the increased flexibility offered by a programmable device. Addition of parameters for new terminal types, or modification of the network configuration and even complete software modules to handle new protocols are possible in an FEP. These features are essential in the dynamic world of data communications.

When designing these devices, most manufacturers choose to emulate their hardwired devices to minimise the problems of conversion for their customers. Passive FEPs are the most commonly used type at present and are perfectly adequate in many environments. They are generally designed in such a way as to give the user the capability to develop his system in phased steps and to upgrade to an active FEP when necessary.

Active Front-End Processors

The value of a front-end processor becomes much greater if it performs some of the functions necessary to control the communications network and terminals as well as carrying out communications tasks. In such devices, known as active FEPs, communications handling is at the user message level between mainframe and FEP, and communications 'knowledge' as well as load is generally removed from the host computer. The CCP is removed from the host to the FEP, while the host retains enough software to handle the FEP itself as a fast peripheral. However this 'network' software, as it is frequently called, can itself be substantial. Figure 7.2 shows a typical configuration containing an FEP.

An active FEP will perform the following functions:

- those performed by a hardwired CCU such as line control and character assembly, recognition of control characters, etc;
- message assembly; the host should receive only complete messages;
- data conversion; data transmission codes are converted into the native data code of the host;
- polling;

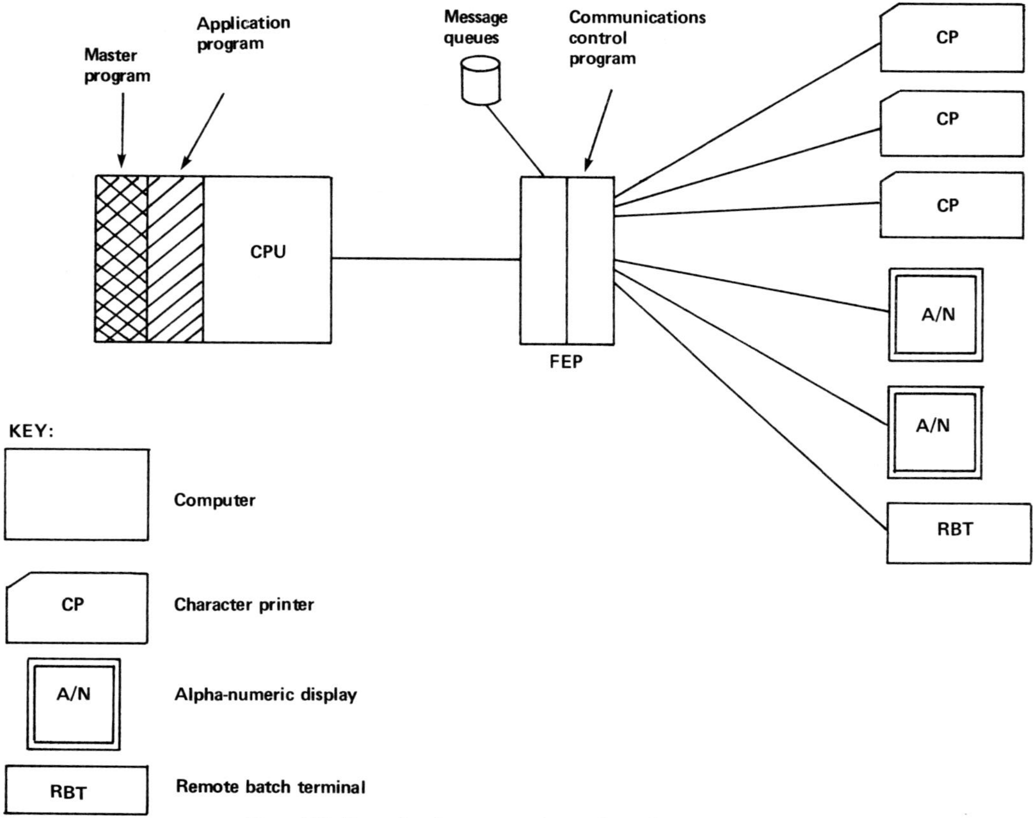

Figure 7.2 Example of a system using an 'active' front-end processor (FEP)

- error control; both hardware and software validation of data is possible;
- message buffering and queueing so that the host is interrupted as little as possible;
- data and message editing;
- message answering; certain simple messages may be handled by the FEP alone;
- message recording; for example, journal records or simply as back-up information to assist in a restart if the host fails;
- removes all concept of lines from the mainframe. The mainframe merely communicates with a number of terminals and knows nothing of the physical topology, eg the number of lines, multi-dropping clustering, etc;
- statistics recording;
- message switching; if the FEP is attached to more than one host it identifies which a message is addressed to and routes it accordingly.

ADVANTAGES OF AN ACTIVE FRONT-END PROCESSOR

An active FEP offers a user several advantages over a passive FEP or a hardwired CCU, but at an increased cost. These are:

- one FEP can handle as many lines as two or more hardwired CCUs and can also serve more than one host. This, together with its expandibility and flexibility, makes an active FEP an economical device as its useful life will generally be longer than a hardwired CCU. Passive FEPs can however usually be upgraded to active FEPs because they are based on the same hardware;
- being an intelligent device an active FEP can be programmed to ensure graceful degradation of the data communications system when the host fails. Recovery from such a failure is much easier if queueing is handled by the FEP as less data is lost;
- the user programs can address the data communications network as a standard peripheral device;
- an active FEP is capable of independent processing with its own set of peripherals;
- diagnostic software can be incorporated in the FEP to help with control and maintenance of the network;

– theoretically a substantial amount of main storage is saved in the mainframe, which may be used for other tasks. This is because the communications control program now resides in the FEP. However network control software has been developed which resides in the mainframe, communicates with the FEP, and performs sophisticated control tasks. This software is frequently far more complex than the CCP it replaces.

LEVEL OF SOPHISTICATION REQUIRED IN A FRONT-END PROCESSOR

When setting up a data communications system for the first time or when adding to an existing one, the user must consider the load that this will place on his mainframe processor. This load can be calculated in terms of the volume and type of traffic to and from each terminal when in use, how frequently it will be used and how many terminals will be used concurrently. Response time to terminal users will fall off as the load increases and the possibilities of getting any other work done while the on-line system is in use becomes scarce. Most mainframes are in fact unable to handle more than a small number of terminals via simple interface units and the point at which some sort of front-end processor is required is quickly reached. The user should consult the mainframe supplier on this as it depends very much on the type and size of the mainframe and the operating system running in it. If line interface units are all that is required these must be obtained from the mainframe supplier. The user must specify the protocols and line speeds required but no further evaluation is necessary.

Having decided that a communications control unit or front-end processor is necessary, the choice now lies between a hard-wired CCU, a software emulation, or an active FEP. Few manufacturers still market hardwired CCUs and a software emulation would almost certainly be more cost-effective; the selection usually comes down to selecting an FEP. Some degree of customising is usual in FEPs, although this normally involves a standard process at system generation time rather than by writing/debugging new code. The cost and value of an FEP can only be assessed in terms of specific situations and systems so it is difficult here to give guidelines on what type of FEP is most appropriate. The importance of the on-line applications is one factor; for instance is the mainframe to be running 90% on-line work and 10% batch or vice versa or a more even split? In general however the sophistication of the FEP should bear some relationship to the complexity of the data communications network it is to support, or perhaps more importantly that which it will have to support in 2 or 3 years' time.

SELECTION OF A FRONT-END PROCESSOR

Figure 7.3 lists the criteria on which a front-end processor should be judged. The specific criteria are discussed individually below. Where no scoring scheme is suggested, as with modems, multiplexers and concentrators, a score of 10 should be awarded to the item that best satisfies the desirable criterion, and lesser scores should be awarded to other items. As with the selection of data communications equipment, it is advisable to visit existing users of an FEP and discuss the performance of both hardware and software before making a final selection.

Supplier

There are three sources of front-end processors – the major computer manufacturers, the plug-compatible manufacturers and the independent systems houses. Because of the close relationship between an FEP and a host, the wisest move for an inexperienced data communications user in this case is to stay with the mainframe supplier. Having said this, two factors – economy and the need for any special features – might suggest otherwise.

The plug-compatible companies and independent systems houses do indeed offer cheaper devices or better cost/performance ratios than the mainframe suppliers. In the late sixties and early seventies there was a large PCM market in hardwired CCUs for this reason. Today however, the market is small. This is primarily due to the spread of intelligence throughout a network – in the FEP, in a concentrator or multiplexer, and even in the terminal. The distribution of functions between these devices can be modified by the manufacturers and unless provision for such an eventuality is made, buying PCM could well turn out to be more costly and certainly more trouble than originally envisaged. This applies primarily to active FEPs.

Such risks may however be worth taking if a user requires some features which the mainframe manufacturer cannot supply. It is in this area that the independent systems houses can be of particular assistance. It must be emphasised however that acquiring an active FEP with non-standard features could well have long-term implications, even outside the telecommunications environment, and

Figure 7.3 Criteria on which to select a front-end processor

*	Supplier
***	Turnkey system
***	Modifications to existing hardware
*	Modifications to existing software
*	Communications software
*	Ability to upgrade device
***	Number of stations pollable on a line
***	Number of host connections
***	Terminal protocols supported
***	Line capacity
*	Peripherals
*	Special features
*	Future enhancements
*	Delivery time
*	Simplicity of installation
*	Reliability, mean time between failure
*	Maintenance support
***	Post Office regulations
*	Supplier factors
*	Costs
***	Compatibility and standards requirements

*** = Essential criterion always

should not be taken without a full appreciation of all the factors involved.

If the user decides to stay with the mainframe supplier then the remaining decision process is generally simple as the choice offered will not be great. The user should however follow through the evaluation process to ensure selection of the most suitable model.

Turnkey System

One of the most important words when acquiring an FEP is 'turnkey'. If an FEP is supplied on a turnkey basis then the user is saved money, time and trouble, assuming no responsibility for the system until it is operating satisfactorily. As all FEPs require some degree of customising, and the software effort involved in installing one could be considerable, installation of an active FEP is a most complex task. Most suppliers will be prepared to work on a turnkey basis. It is recommended that this criterion is one of the most important on which a supplier should be judged.

Modifications to Existing Hardware

Any FEP should require no modifications to existing hardware and should be completely compatible with it. This, like the need for a turnkey system, is extremely important and should be considered an essential criterion.

Modifications to Existing Software

Carrying out no modifications to existing software is an extremely desirable criterion but often cannot be met unless the mainframe supplier is supplying the FEP. Changing from line or data adaptors to a hardwired CCU or one of its software emulations will require additions or modifications to mainframe communications software, while changing to an active FEP, will require considerable redistribution of mainframe communications software as well as modifications to applications programs. Such changes are inevitable but the user must at all costs ensure no changes are necessary to the operating system or other mainframe software so that it is no longer compatible with the mainframe manufacturers products. The suggested scoring system is:

No modifications to existing software	10
Modifications to existing software but still compatible with mainframe manufacturers products	5
Modifications to existing software but possible incompatibility problems	1

Communications Software

A user implementing a data communications system for the first time may need special communications software, while a user upgrading his system with a front-end processor may need to acquire new software and modify some existing software. Installation of an active FEP can be much more complex than installation of any other network interfacing device because of the work required on software. Some of the software is generalised and some highly specialised, relating to the specific network the FEP is to support and the applications it is to be used for. The mainframe supplier must be consulted on this as early as possible as software installation can be a long and complex process.

Ability to Upgrade Device

For many users taking their first steps in data communications a passive FEP is the most suitable device. However, it is highly likely that an organisation will wish, or will be forced by the mainframe supplier, to upgrade to a more sophisticated device in the future. This step will be facilitated if the passive FEP can be upgraded, ie if the active FEP is based on the same hardware and software changes only are required.

An example of this is the IBM 3704 and 3705 FEPs. Either of these can be used as a passive FEP (this is known as EP mode), or as an active FEP (NCP mode). The transition requires considerable software modifications in both mainframe and FEP, but the basic hardware remains unchanged. The transition can be undertaken gradually, some software still interacting with the FEP in EP mode while other software has been upgraded to NCP mode.

This criterion should be considered carefully in evaluating an FEP as it can result in considerable economy in the long run. It could be viewed as essential in some organisations while others may consider it desirable, scoring 10 for an upgradable device and 1 otherwise.

Number of Stations Pollable on a Line

Some FEPs place a restriction on the number of stations that can be polled on a line. The user must ensure that this number is adequate for his present network and its anticipated growth. If the network configuration requires a large number of stations on any line, at or near the limit of the FEPs under consideration, then it is advisable to consider redesigning this section of the network; the response time to the users may well become unacceptably low and if a line failure occurs many terminals will be disabled. The ability to satisfy the number of stations pollable on a line will be an essential criterion where such a configuration is to be used.

Number of Host Connections

Active and passive FEPs, unlike CCUs, may be attached to more than one host. This feature may well be of importance to a user with more than one mainframe (not necessarily at the same site), or a user whose mainframe is approaching the limit of its capacity and may well be augmented with another in the foreseeable future. This requirement to support N host connections, where N is the number of host connections that the user anticipates he will require, will be an essential criterion.

Terminal Protocols Supported

The question of protocols supported is particularly relevant to a hardwired CCU or an emulation; an active FEP can be programmed to deal with any protocol although if it is an uncommon or special one, the cost of providing such support may be an extra. The most commonly supported protocols are asynchronous teletype, BSC, HDLC and SDLC. Most FEPs will provide for a mix of these. This criterion is an essential one as by this stage the line protocols, terminals and other equipment which is protocol dependent will be in the process of evaluation as well as the FEP. Thus a change in protocol could entail considerable additional work.

It may also be useful to introduce a desirable criterion that all necessary protocols are supported for the minimum additional cost. This criterion could be scored in such a way that 10 points would be awarded if no additional cost would be incurred, then lower scores in relation to the degree of extra cost.

Line Capacity

FEPs and CCUs vary in the total number of lines they can deal with, this total depending on whether the lines are full- or half-duplex. This number is generally quite high; for example, an ICL communications link controller has a capacity of 128 full-duplex lines. Some devices are modular, giving the user the option of upgrading, from 56 to 128 lines, for instance, at a later date as the network grows. This ability to upgrade can be a useful feature as the user does not have to pay for a large capacity device initially when much of the capacity is not required until a later stage in the development of the data communications network.

Users are not usually limited by the capacity of the CCU or FEP but it is nevertheless important that this factor is taken into account in the evaluation exercise. The device must meet capacity requirements as an essential criterion. However the evaluation is complicated by the issue of upgradability and how large the network is likely to grow in the foreseeable future. Hence a desirable criterion may also be necessary. A suggested scheme if there are no plans for the network to grow over the next three to four years is:

Device meets capacity requirements and does not exceed them excessively	10
Device meets capacity requirements but exceeds them excessively	6

If the network is expected to grow, an alternative scheme could be:

Device meets capacity requirements, does not exceed them excessively and is upgradable	10
Device only just meets capacity requirements but is upgradable	6
Device meets capacity requirements, exceeds them excessively and is not upgradable	4

Peripherals

Some active FEPs can support peripherals and function as independent processors. This fact may well be important to some organisations, in which case it should be viewed as an essential criterion. Other organisations may consider it desirable but not essential, scoring 10 for an FEP which supports a range of peripherals and 1 for an FEP which supports only a console or no peripherals.

Other Features

FEPs and CCUs are available with a range of sophistication, some incorporating particular features not commonly available such as the ability to broadcast messages when the host is out of action. Such features are often the reason why an organisation selects a particular device over all others. If a special feature is vital then economics do not enter into it, more usually however a feature may be attractive but not essential. In this latter case the organisation must decide the worth of the feature to itself and make the final decision accordingly.

PCM manufacturers in particular supply FEPs with additional

features while mainframe suppliers will sometimes supply such features in software packages for an additional cost.

Selecting an active FEP with non-standard features may well have long-term implications for the computer system as well as the data network. It may, for instance, mean that eventually it may be necessary to run unsupported mainframe software. Further non-standard features in the FEP need non-standard software in the mainframe to support them. Although many people have taken this step it is extremely risky and not advisable.

Future Enhancements

When acquiring an FEP, especially an active one, it is important to consider the supplier's attitude to a new development in software. Will the supplier continue to enhance the software to remain compatible with and to exploit any new features in the mainframe software? How long will he take to do so? If the FEP is not from your mainframe supplier it is important to consider if such enhancements are indeed possible.

8 Switching and Patching Equipment and Racks

INTRODUCTION

Switching and patching equipment both serve the same purpose although they function slightly differently. They are installed to make life easier for the network operators. Without this equipment, switching lines and data communications devices with all the cabling involved, can be a long, complicated and untidy (if not entangling) job.

Racks are another piece of equipment used in a network for convenience. A rack is basically a cabinet with a power supply which will house a range of modems, multiplexers and other data communications equipment.

SWITCHING AND PATCHING

Switching and patching are two ways of performing the same function – disconnecting one path from a device for data transmission and establishing another. This can be performed without such equipment, though less conveniently.

Patching equipment consists of a card with a plug at each end and two sockets. Lines or V24 interface cables are connected to the sockets so that plugging the card into the sockets generally establishes a connection between them. An alternative method is for two sockets to be connected internally, the insertion of the cable then breaking the existing connection and allowing one or both sockets to be patched to others for different configurations.

Switching equipment permits a different path to be opened by use of a switch instead of a cord. In addition equipment combining both switching and patching is available. Patching equipment is the most commonly used as it is more flexible, allowing any socket to be connected to any other, and is generally cheaper. The choice between the two will depend on the network's specific operational requirements and size. Table 8.1 gives an indication

Number of modems	Switching/patching requirements
2	No switching equipment is really necessary although it could be used to make life easier for the operator and reduce the possibility of error/damage
5, 6 +	Switching is necessary to help the operator and keep all cabling neat
30 +	Remote switching/centralised network management should be considered. Cost justification will depend on availability of staff, both centrally and at the remote sites and the degree of urgency in data traffic flows

Table 8.1 Switching and patching requirements

of the general switching and patching requirements in relation to network size.

Switching and Patching Points in the Network

There are three points in the network where switching or patching can be implemented:

– V24 interface (fig. 8.1)

  This is at two points on a communications link: between the modem (DCE) and the terminal (DTE), where most switching and patching takes place; between the modem and the computer port; or between the modem and a front-end processor or multiplexer at the computer.

– Line (fig. 8.2)

  This is on the analogue side of the modem at the point where the modem is connected to the Post Office circuit.

– Host computer site

  If there is a network control system at the central site of the network, switching of computer channels can take place there.

Functions

Switching and patching equipment is used in three situations:

– fallback/standby;

– testing and monitoring;

– network reconfiguration.

Fallback/standby

Switching or patching equipment can be used when either the line or a data communications device fails. If, for instance, a private line fails, there may be fallback to the PSTN or there may be a standby private circuit. In either situation the equipment is used to disconnect the modem from the failed line and connect it to the standby line.

Similarly, if the modem, or another hardware component in the network, should fail and a standby device is available, switching/patching enables the spare equipment to be used. The line is disconnected from the failed device and reconnected to the spare equipment.

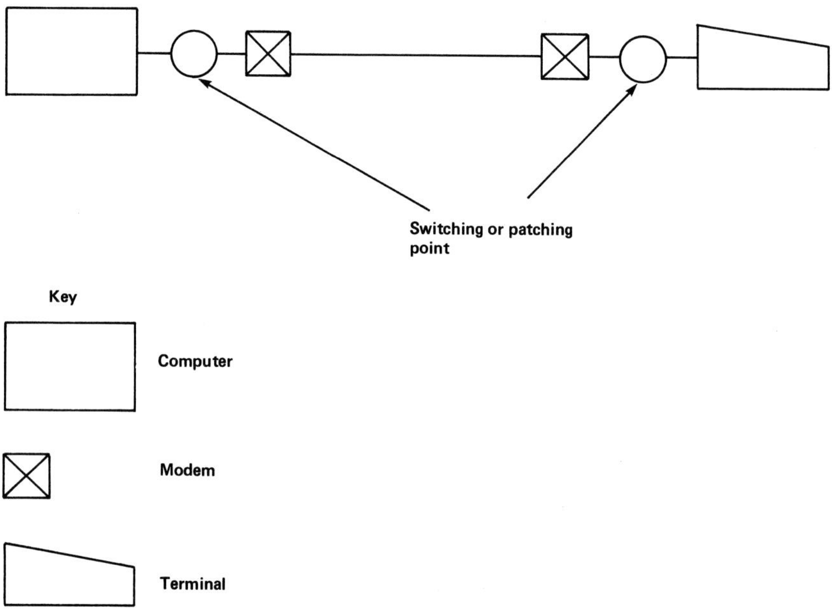

Figure 8.1 Switching and patching points — V24 interface

(a) No front-end processor

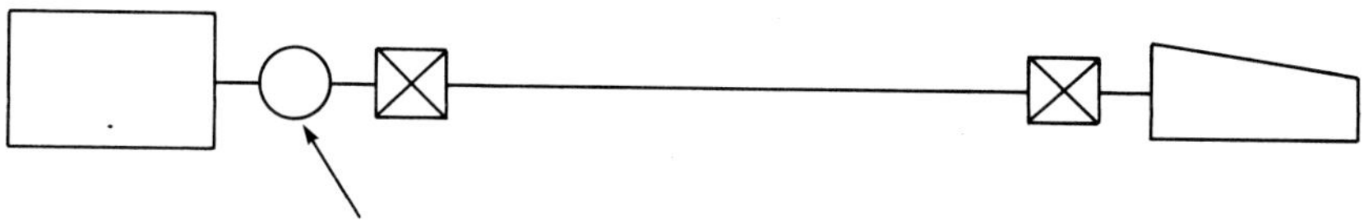

(b) With a front-end processor

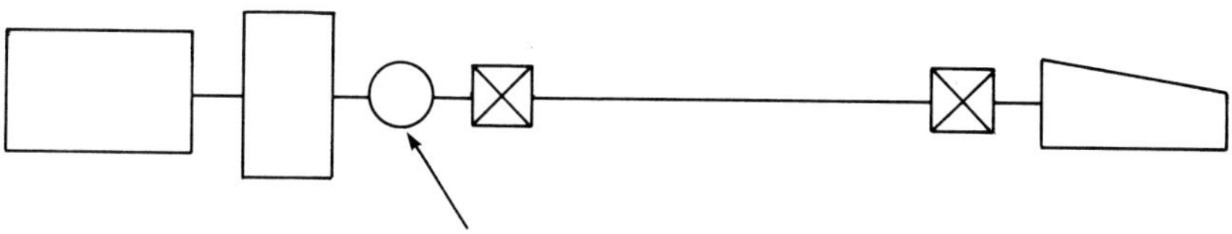

Figure 8.2 Switching and patching points — modem and computer port

Testing and monitoring

Switching/patching equipment can assist in testing both circuits and equipment. To isolate a fault, the circuit and equipment can be switched/patched off-line. Tests and monitoring activity can then be carried out on each component using either the equipment's loopback/self test facilities or external looping facilities. Alternatively special test equipment can be switched/patched into the network. Testing is normally initiated at the host computer end, eg by a network controller.

Network reconfiguration

Switching/patching equipment may be used if the network is reconfigured in any way. It makes the reconfiguration easier and also reduces the possibility of mistakes or damage to equipment.

Reconfiguration may be necessary if there is equipment or line failure, if the network changes in design or if the network expands. It is especially useful if permanent connections are not required for example, one particular terminal is to be used during the day and a different terminal used for processing during the night.

As a consequence of these functions, the most common use of switching/patching equipment is in the following ways:

- modem to spare port connection (fig. 8.3);
- fallback line connection. For example the operator can switch or reroute a V24 interface signal from normal to fallback mode. However with Post Office modems there will usually be a facility which switches between the private line and the PSTN. On the other hand, with privately-supplied modems, the manufacturer usually supplies a line adaptor unit which can effect the changeover (fig. 8.4);
- line to spare modem connection (fig. 8.5);
- test set connection, usually at the V24 interface;
- central site switching at the operators console;
- modem to standby computer, front-end processor, or multiplexer (fig. 8.6).

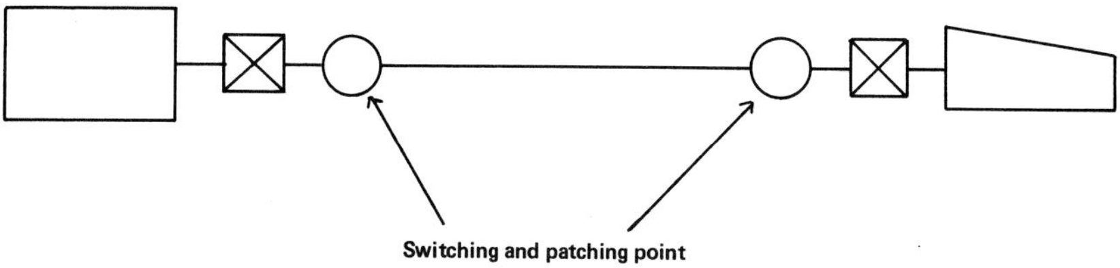

Figure 8.3 Switching and patching points — line

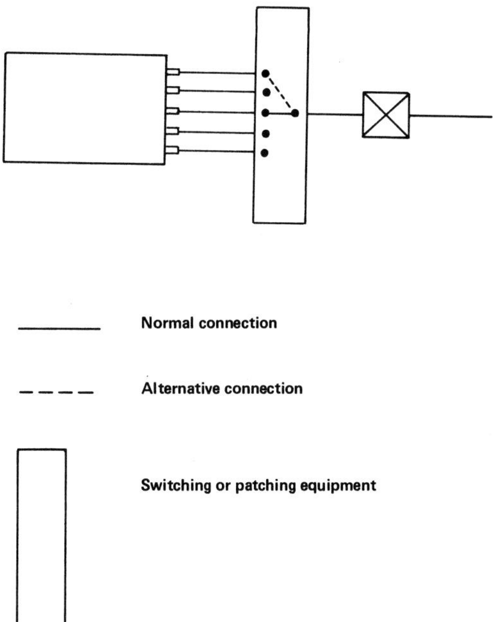

Figure 8.4 Modem to spare port connection

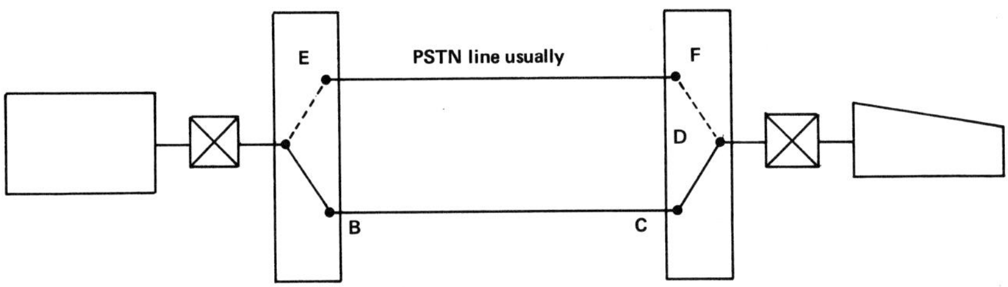

Key

A B C D Normal connection

A E F D Alternative connection

Figure 8.5 Fallback connection

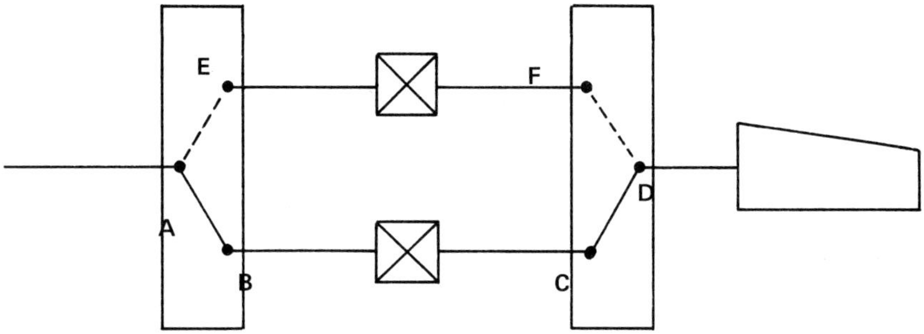

Key

A B C D Normal connection

A E F D Alternative connection

Figure 8.6 Line to spare modem connection

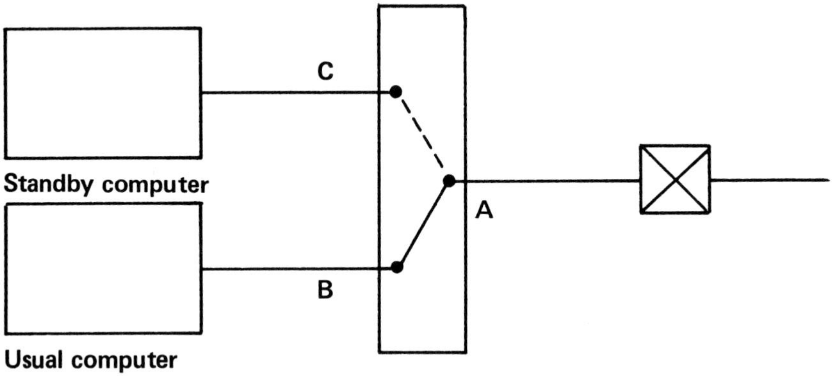

Key

A B	Normal path
A C	Alternative path

Figure 8.7 Modem to standby computer

SELECTION OF SWITCHING AND PATCHING EQUIPMENT

The basic switching/patching function is straightforward and this is reflected in the small range of features available on switching and patching equipment. These features are discussed below. The only feature which is always an essential criterion is the number of lines or circuits which the equipment can handle. This must be adequate to meet the network's requirements. The specific and general criteria are summarised in Figure 8.7. Unless otherwise stated, each criterion can be scored by awarding 10 to the most satisfactory model, and lower scores to others, according to how closely they satisfy the criterion.

Features

– capacity

Equipment is generally modular in construction. As the network expands and the number of channels (modems) grows, the capacity of the equipment can be increased by adding on modules. Modules can be purchased which provide switching for groups of 8 or 16 separate channels as well as for single lines. Table 8.2 shows the cost of various capacities of switching/patching equipment. It is essential that the equipment satisfies the required capacity.

– local/central operation

Switching but not patching equipment at remote sites may be controlled remotely by an operator at a central site. This requires sophisticated and therefore more costly switching equipment.

– size

The size of equipment needs to be considered if space is limited for it varies greatly. Usually equipment is designed to fit into the standard equipment size of 17″ wide and is usually between 3″ and 7″ in height.

– mounting of equipment

Equipment can be purchased to connect into the network in a number of different ways.

- stand-alone with no kind of mounting at all, maybe fitting into the operator's desk if one exists;
- secured in a central console panel;
- fitted into a rack along with the modems.

Number of channels/lines	Cost (approximate)
1	£ 50 – £ 130
2	£ 70 – £ 180
4	£150 – £ 250
8	£300 – £ 550
16	£400 – £1000

Note: Switching of lines is cheaper than switching at the V24 interface. Remote site switching initiated at central site is more expensive than local switching operation

Table 8.2 Switching/patching equipment costs

– design

Switching and patching should be designed in such a way that it is easy to use. For example, it may be useful for all channels to be able to be switched simultaneously as well as individually. Also with patching equipment a straight-through link eliminates the need for patchcords or looping plugs for standard connections, giving a much tidier effect.

Equipment Selection

In deciding what equipment is necessary for the network under consideration, various factors must be considered.

- What does the network configuration require? For instance if no standby equipment is to be held then switching/patching equipment will not be necessary for this function.
- How is the network planned to expand and will the switching system be capable of expansion or modification to match this?
- Who is to operate the equipment? The main aim of switching/patching equipment is to facilitate operation of the network, and to aid its operators. The equipment the user selects should be easy for the operator to understand and operate. Staff may require training to use the equipment.
- What testing, monitoring and reconfiguring equipment already exists in the network? Many modems have looping and other self-test facilities. The network may also have multiplexing or concentrating equipment with a variety of built-in diagnostic features. These features may well preclude the necessity for switching and patching at some points in the network.

RACKS

Like switching and patching equipment, racks are not an essential part of a data communications network. However the larger the network, and hence the more equipment required, the more useful they become. A rack is basically a cabinet, usually with a power supply, which holds data communications equipment such as modems, multiplexers, test equipment and switching and patching equipment. The power supply can be common to all the equipment held in the rack, which sometimes will have in addition, a standby power supply.

Figure 8.8 Criteria for the selection of switching and patching equipment

*** Capacity
* Local or central operation
* Size
* Mounting
* Design
* Cost
* Simplicity of installation
* Delivery time
* Reliability
* Maintenance
* Supplier factors
*** Compatibility and standards
*** Post Office regulations

*** = Essential criteria always

Racks are custom-built, usually by suppliers of the equipment they are designed to hold and various diagnostic features can be built in if required.

9 Test and Diagnostic Equipment

INTRODUCTION

There is a wide variety of communications test equipment available today, ranging from microcomputer- or minicomputer-based network control systems costing tens of thousands of pounds to simple units for testing one transmission characteristic costing around a hundred pounds. When setting up a data communications network an organisation is faced with deciding how much (or how little) it can afford to spend on such equipment. No hard and fast rules can be derived, and an organisation must base its requirements on the importance of the network to its functions. If the network is vital then substantial investment is required in network control equipment, standby equipment and in personnel. Even if network failure is not catastrophic a certain basic set of hardware is necessary to give the user the ability to identify the faulty item and arrange maintenance from the correct supplier. This results in a better supplier/user relationship and generally gets the fault repaired more quickly.

No matter how much is invested in equipment for testing and diagnosis, little will be achieved unless there is someone in the organisation with sufficient understanding of data communications to use it. Much of the equipment on the market today is extremely sophisticated and requires a skilled person to operate it and interpret the results.

The aim of this chapter is to describe the basic equipment necessary, to assist the reader in its selection, and to describe the range of additional equipment on the market and the circumstances in which this may prove of use. The subject is complicated by the fact that many test sets are capable of carrying out more than one test, while the grouping of tests into multi-test units is not consistent between manufacturers. The tests can be classified in the following way:

- transmission tests;
- data tests;
- protocol tests.

In general, a multitest unit is only capable of carrying out tests within one of these classifications. Tables 9.1.1 and 9.1.2, reproduced from *Fault Diagnosis in Data Communications* (NCC Publications), give examples of the various test sets on the market and the tests they are able to perform.

TESTING AND DIAGNOSIS

Before discussing the available test equipment, it is worthwhile to consider exactly how testing and diagnosis should proceed. Unless it is approached in a logical manner, no amount of expensive equipment will help the user to isolate and correct the fault quickly and efficiently. There are three elements in this process:

- procedures;
- a suitably trained network or communications controller;
- a good relationship with suppliers.

Procedures

While the network is being planned and implemented, fault-finding procedures should also be drawn up. This should include:

- defining areas of responsibility;
- ensuring good documentation is available to assist all users of the network in fault-finding;
- ensuring all users of the network are able to help when necessary in fault-finding;
- establishment of a reporting procedure for faults so that troublesome areas can be identified. For example, if frequent errors occur on a line, this can be an indication of deterioration in the line quality and it can be corrected before total failure occurs if spotted in time;
- establishing a fault-logging scheme and a mechanism for bringing important or long-standing faults to the attention of higher management;
- ensuring that good voice communications facilities, not dependent on the data communications system, are available between sites. When faults occur, fault-finding procedures

Equipment Type	Transmission Testers							Data Test Equipment								Protocol		
	Signal Generator	Level Measuring	Group Delay	Noise Level Measuring	Impulse Noise Counter	Audio Monitoring	Frequency Measuring	Pseudo-random – Synchronous	Pseudo-random – Asynchronous	Interface Modem Monitor	Interface Through Monitor	Multiplexer Test Set	Fixed Pattern Test Set	Polling Test Set	Distortion Measuring	Protocol Analyser	Protocol Analyser and Store	Message Generator
Borer																		
Distortion Measuring Set DMS4															✓			
TMG 10 Text Message Generator									✓				✓					
Data Tech. 9600 DTS-1								✓	✓	✓	✓				✓			✓
Intershake II DTM-2										✓	✓	✓		✓		✓		✓
CASE																		
270 Data Test Set											✓	✓	✓					✓
Cole Electronics																		
IDS Model 60 Interface Monitor											✓	✓						
General Audio and Data Communications																		
Datascope D-601 B											✓	✓			✓	✓		
High Speed Tape Unit T-611											✓	✓			✓	✓		
International Aeradio																		
PMDS Monitor											✓	✓						
Marconi Instruments																		
Data Line Analyser TF 2809	✓	✓		✓														
Nolton Communications																		
1608 Asynchronous Data Terminal Tester											✓	✓			✓	✓	✓	
1609 Synchronous Data Terminal Tester											✓	✓			✓	✓	✓	
Racal-Milgo																		
220 Transmission Test Set						✓		✓	✓				✓		✓			
330 Transmission Test Set								✓		✓								
340 Transmission Test Set								✓	✓	✓								
711 Dynamic Test Set														✓				
Rank-Hovis-MacDougall																		
DM300 Data Monitor										✓	✓				✓	✓		

Table 9.1.1 Test equipment
Source : Fault Diagnosis in Data Communications

Equipment Type	Transmission Testers							Data Test Equipment								Protocol		
	Signal Generator	Level Measuring	Group Delay	Noise Level Measuring	Impulse Noise Counter	Audio Monitoring	Frequency Measuring	Pseudo-random – Synchronous	Pseudo-random – Asynchronous	Interface Modem Monitor	Interface Through Monitor	Multiplexer Test Set	Fixed Pattern Test Set	Polling Test Set	Distortion Measuring	Protocol Analyser	Protocol Analyser and Store	Message Generator
Siemens																		
Data Test Signal Generator 7H2																		√
Telegraph Distortion Meter 2H1															√			
Data Test Set 2H2								√	√									
Impulse Noise Counter 4H1					√													
Level Oscillator W330	√																	
Level Meter D349		√		√														
Systron Donner																		
225-3 TDM/Modem Test Set								√	√	√		√						
223 Pseudo Noise Transmission Test Set								√										
224 Pseudo Noise Transmission Test Set								√	√									
Technel Data Products																		
MS12 Data System Monitor								√		√	√							
Trend																		
Data Transmission Test Set 1–8								√	√	√			√	√	√			
High Speed Data Transmission Test Set 5 MKII								√					√		√			
Data Transmission Test Set 1–4								√	√	√			√		√			
Telegraph & Data Signal Analyser															√			
Telegraph & Data Message Generator																		√
Venner																		
Impulse Noise Counter					√													
Wayne Kerr																		
44C Transmission Measuring Set	√	√																
36A Selective Measuring Set		√					√											
2B Psophometer				√														
35A Portable Level Measuring Set		√																
Wandel & Goltermann																		
SPM-3 Level Meter		√		√			√											
PS-3 Signal Generator	√																	
LD-30 Group Delay	√	√	√				√											

Table 9.1.2 Test equipment
Source : Fault Diagnosis in Data Communications

can then be coordinated easily as operators and users at different sites can still communicate.

Communications Controller

No matter what the size of the network, someone in the organisation must have sufficient understanding of the principles of data communications to be able to undertake overall responsibility for fault-finding and liaison with suppliers. A suggested job description for such a person, the communications controller, is given in *Fault Diagnosis in Data Communications Systems* (NCC Publications).

The job may not be a full-time one but when a fault occurs, the communications controller must be able to give priority to identifying and rectifying it. He must be able to use any test equipment bought for the network as well as use any test facilities built into hardware such as modems. No matter how much the organisation has decided to rely on its suppliers when a fault occurs a certain amount of basic checking must be done before a supplier is called. For instance, the communications controller should:

- check that the power supply is functioning;
- check that all the plugs in any patching or switching panels are in the correct positions;
- check that the computer is not down;
- check that the switchboard has not failed if any PSTN lines are used;
- check with any remote user who has reported a failure that his modem is correctly connected to the terminal;
- check for any user errors at the remote site;
- check for any operator errors at the centre;
- gather as much information on the fault as possible, whether by the use of test equipment or not.

In addition to such checks, some suppliers, especially the Post Office, require someone in the organisation to be available to assist in testing.

Relationship with Suppliers

A good relationship with equipment suppliers can only be beneficial. One person, generally the communications controller, should be responsible for supplier liaison, even for equipment at different sites. In a situation where more than one supplier is involved, agreed procedures for dealing with faults at supplier interfaces should be established before the network is operational, to avoid disputes later. In all cases, the more information the communications controller can give the supplier about the fault the better.

BASIC EQUIPMENT

If the network consists of a few point-to-point or multipoint links using only Post Office equipment and is not vital to the business then an organisation can get away without any test equipment. For instance a system used purely for program development by research scientists can perhaps be out of action for a day without any serious consequences. In all other circumstances, however, especially if more than one supplier is involved, a basic set of test equipment is essential to any testing of a data communications network. This basic equipment is:

- a data test set;
- facilities for line looping;
- a high impedence earpiece;
- a tool kit (an obvious but frequently forgotten requirement).

This equipment can be used by a person (with a thorough understanding of the network configuration but only a basic knowledge of the principles of data communications) to identify the probable source of a fault in a network.

Data Test Set

A data test set, alternatively called a modem test set, is a device which performs a test or a number of tests which measure the ability of the communications system to transmit digital data. Ideally two sets are required so that a line can be tested from both ends. For the basic equipment for testing and diagnosis of a point-to-point network two basic data tests are essential. These can be found in one piece of equipment or separate data sets can be bought. Any other tests can be useful but are not essential. Slightly different features are necessary for a network with multiplexers or multipoint links. Data test sets vary in price from £200 to £1000 depending on the number and complexity of the tests they perform.

Features Important in the Selection of a Data Test Set

Figure 9.1 lists the features which should be looked for in a data test set. Those features marked with asterisk must be considered for every data test set while those unmarked are tests which may be found alone in a single test set or may be found in various combinations in a multitest set. In any selection exercise a list of criteria must be drawn up which includes all the general criteria as well as the asterisked features in Figure 9.1.

Figure 9.1 Specific criteria for selecting a modem test set

***	Test set matches modem features
***	V24 interface
*	Portable or rack mounted
***	Pseudorandom pattern transmission
*	Fixed pattern transmission
***	Interface monitoring
*	Interface signal response measurement
*	Bias distortion measurement
*	Peak isochronous or telegraph distortion measurement
*	Block error count
*	Multidrop testing
*	Multiplexer testing
*	V35 interface

*** = Essential criterion always

Test Set Matches Modem Features

This is an essential criterion for selection of a data test set. The modem features which must be matched are:

- the presence of an internal or external clock;
- synchronous or asynchronous transmission;
- data transmission rates;
- simplex, half-duplex or full-duplex transmission.

Some test sets are restricted in the options they offer on those features, while others allow variations in all four cases. This variation is extremely important in a network with a variety of modems.

V24 Interface

To be compatible with the modems, a V24 interface on the test set is essential while input and output interchange circuits must be compatible with V28. Care must be taken with equipment from U.S. manufacturers which may conform to the U.S. equivalent of V24/V28, namely RS-232-C. In this case the pin connections may need one or two minor rearrangements to conform exactly to V24/V28, but the equipment should not be dismissed because of this.

Portable or Rack Mounted

Portable test equipment is useful at sites where the modems are scattered around and in networks where there are sites with only one or two modems. In the latter case it may not be worth buying one test set per site while a portable one can be shared between many sites. Rack-mounted modem test sets are suitable at a site where there are many modems grouped closely together. Even in this case, care must be taken as the test set could be mounted in a rack at one end of a room and is required to test a modem at the other: it must then be disconnected from the rack for use.

In general, portable data sets offer greater flexibility. The rule would appear to be, if in doubt get a portable one, as this can always be conveniently stored on a shelf in a rack. Some simple test sets are indeed only available in portable versions.

Pseudo-Random Pattern Transmission

The normal way to test modems and lines is to transmit a known data pattern and check the received data for errors. Normally a fairly short data pattern representing 'random' data is transmitted repeatedly. Three such patterns are in common use, a 63-bit pattern (normally used on telegraph links), a 511-bit pattern (specified in CCITT recommendation V52), and a 2047-bit pattern.

A modem test set usually consists of a transmitting section and a receiving section, which are almost independent. It connects to the V24 interface of the modem (the terminal must be disconnected) and generates, usually under switch control, all the standard interface signals. The line is either looped or a second test set is connected at the other to receive the data stream.

The transmitter in the test set outputs the selected pattern as transmitted data to the modem. The receiver synchronises to the incoming received data stream, then examines the data for the expected 63-, 511- or 2047-bit pattern. When it identifies the pattern, the receiver synchronises its own checking logic with the received data. The test set then displays a count of the number of bits received incorrectly. Such a test, run for approximately twenty minutes in both directions is a good indication of the suitability of lines and modems for data transmission.

This test is an essential feature of a data test set selected as part of a basic test set.

Fixed-Pattern Transmission

In addition to pseudo-random bit patterns, fixed bit patterns can be useful for testing. Some test sets have the ability to transmit a defined pattern such as all 1s, or all 0s, or alternate 1, 0, etc, or other similar fixed-bit patterns. This is not really a separate test but an additional feature on modem test sets which transmit pseudo-random patterns.

Interface Monitoring

The ability to monitor the V24 interface is an essential feature of a modem test set for a basic test set. The feature can be either incorporated in a multitest set or is available as a separate portable device. By checking on the conditions across the V24 interface the monitor will confirm whether the modem is receiving/generating the correct control signals.

Interface Signal Response Measurement

Measurement of the response time of signals, such as the Ready for Sending signal, can confirm that the correct sequence of signals to initiate transmission is being sent from the computer and received at the terminal. This is an effective test when all components of the system check out individually but the overall system does not work.

Bias Distortion Measurement

Bias distortion relates to the timing of the received data elements on the V24 interface. Figure 9.2 shows a signal which has bias distortion. Its measurement can be carried out on some modem test sets and its presence indicates a possible fault in the digital modem. This facility is not necessary in a basic test set. It is included in some multitest modem test sets or alternatively available on a separate device.

Peak Isochronous or Telegraph Distortion Measurement

This distortion of the signal (fig. 9.3), like bias distortion, indicates a fault in the modem. The facility of measuring it is not necessary in a basic test set. It is included in some multitest modem test sets or alternatively is available as a separate device.

Block Error Count

All modem test sets which transmit pseudo-random or fixed test patterns display the number of bits in error in the received pattern. A useful additional feature is a display of the number of bits in error in a block. This is not a separate test but an additional feature of a modem test set. It indicates the effect of transmission errors on data transmission.

Multidrop Testing

To test a multidrop line, a test set is necessary which can simulate polling and controlled carrier systems. Pseudo-random pattern tests determine the relatively long-term performance of a communications system. A polling test complements a pattern test by simulating the polling carrier reversal situation.

Multiplexer Testing

A modem test set used in a network which employs multiplexers must be able to drive the multiplexer inputs as well as the modem, ie generate asynchronous characters and not just a pseudo-random bit stream. Generally the two features are not found in the same test set.

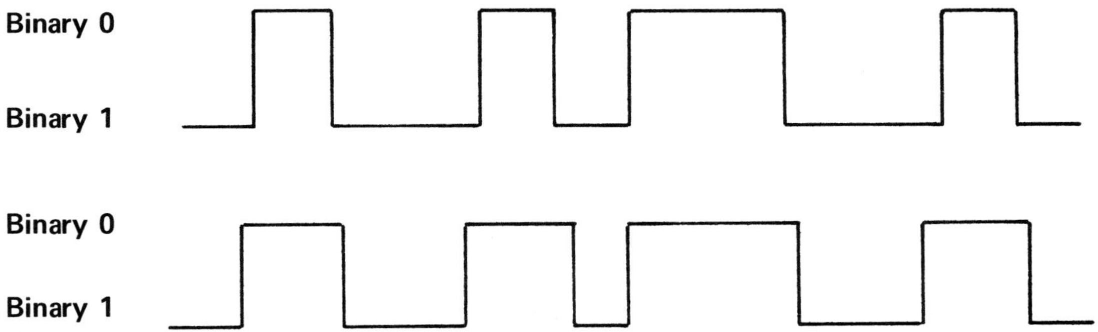

Data with Binary 0 bias distortion

Figure 9.2 Bias distortion

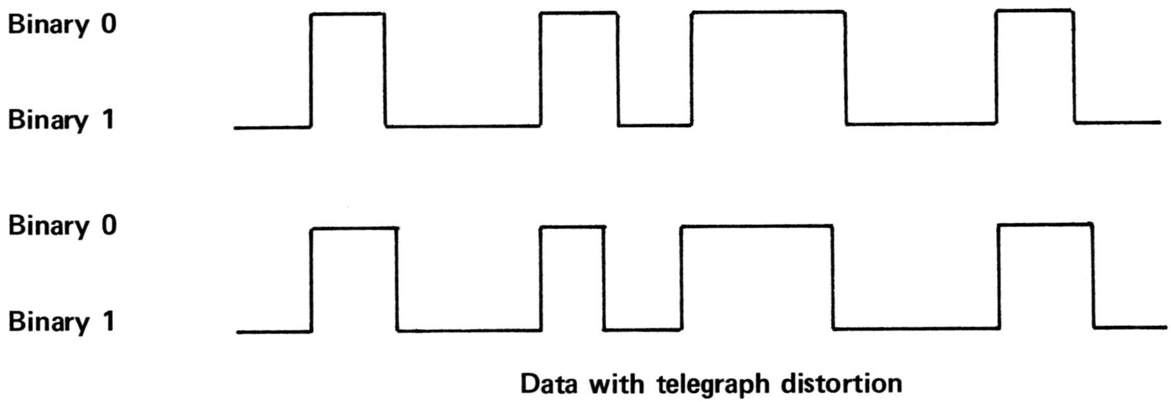

Figure 9.3 Telegraph distortion

V35 Interface

In a network which employs modems or lines operating at greater than 20,000 bps, a test set with a V35 interface is necessary. Such a test set may perform any or all of the tests described above but via the V35 instead of a V24 interface.

Facilities for Line-Looping

Looping of a line so that transmitted data is returned to the point at which it was transmitted via a known path is an extremely useful testing tool. Despite the complexity of a connection between a terminal and a mainframe, each section of the link can be treated as a point-to-point connection.

Figure 9.4 which shows a simple point-to-point link between a terminal and a mainframe. If a test set is inserted at point X and the line looped at A when the modem transmit and receive circuits are tested, the line is then looped at B, testing modem 1 and the line, and then at C, testing both modems and the line. The pattern can be repeated in the opposite direction by inserting a test set between the computer and modem 2. It is important to ensure that the loop is applied exactly at the intended point and that signals are being transmitted and received in the expected places.

To prove that a loop has not been wrongly attached elsewhere on the line, always check that removal of a loop causes a test to fail. In addition to prove that any receiving signal being used is coming from the correct source it is necessary either to remove the source or change its signal in some way.

Looping is not however a complete check especially as the line characteristics may be worse on the loop, which consequently may not work.

It is advisable to ensure that facilities for line looping are available at either side of all items of hardware in the network. The Post Office private circuits are terminated in line termination units, known as Case 200A boxes, which permit looping. A Case 200A unit is a grey plastic box about seven inches square and is normally wall-mounted close to the modem. Where there is more than ten or more circuits the Post Office may install a line termination rack which contains the individual Case 200s rack mounted together.

If there has been a failure, after the user's own equipment has been checked, tasks can be carried out using the looping facilities on this equipment. The user can carry out the tests or may assist

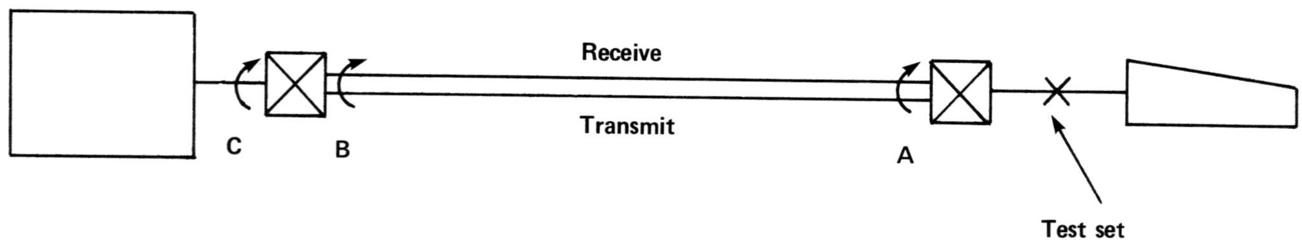

Figure 9.4 Looping on a point-to-point link

in testing the line under instruction from Post Office engineers. Some modems have more looping facilities built in than others but looping can also be achieved by the use of switching and patching panels (as described in Chapter 8). Looping can be done automatically in a network with a sophisticated network control system, but more usually it requires a person to move some plugs or throw a switch.

A High Impedence Earphone

This can be used to listen to a line without interfering with transmission, although it may cause line errors when being connected or disconnected. On a private line a voice adaptor on the modem will enable the user to find out if the line is operational simply by talking down it. This must be done with the terminal disconnected, as shown in Figure 9.5.

However use of an earpiece between the modem and the line termination unit will enable the user to listen to the signal on the line. This test can convey a great deal of information about the state of the circuit, although experience is required to identify and analyse the tones heard. The earpiece and an attaching point can be obtained from the Post Office on request. The Post Office also have training tapes illustrating the various tones employed by the Datel services.

OTHER TEST EQUIPMENT

Transmission Test Sets

Transmission test sets are designed to test the various characteristics of the transmission circuit to see if it is within the Post Office specification. This of course only applies to leased circuits. Such tests include measurement of the following transmission parameters:

- amplitude versus frequency response;
- group delay versus frequency response;
- psophometric noise measurement;
- impulse noise measurement;
- quantisation noise measurement on PCM links.

Most test sets include a tone generator which produces signal of known frequency and power level which is useful for continuity checking. The transmission parameters are also measured with

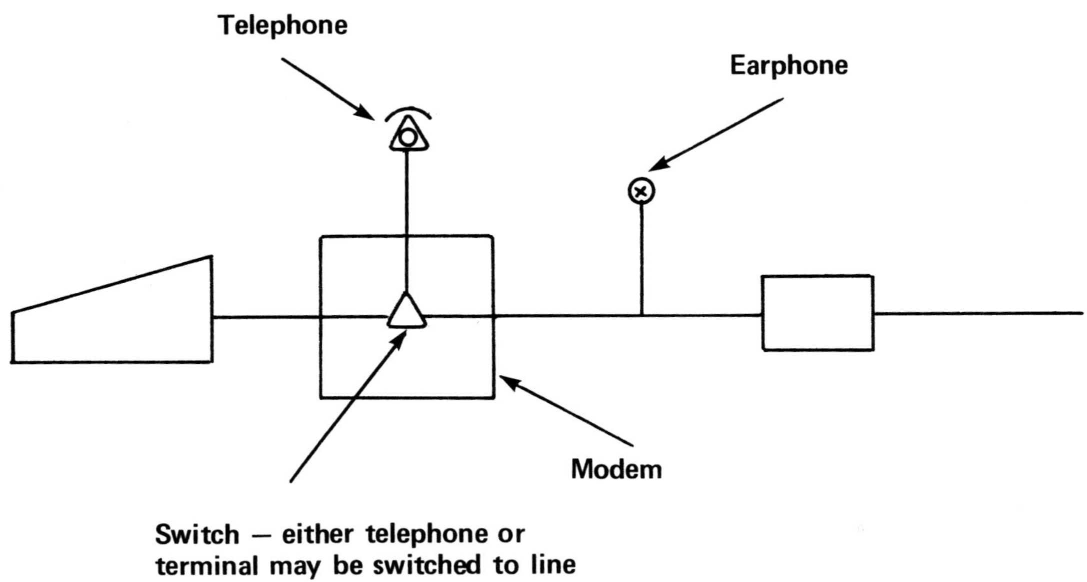

Figure 9.5 Using an earphone on a line

regard to this signal. Many modems can provide a suitable signal for testing however. Transmission test sets can be useful for deciding if a fault lies in a modem on the line when total failure has not occurred but are not essential testing tools.

In selecting a transmission test set a list of the desirable features can be drawn up and each model evaluated given a score of 10 if the feature is present and 1 if not.

Protocol Test Sets

This equipment is used to analyse actual line traffic and to determine whether the protocol on a link is being observed by its various components. Such equipment, fairly new on the market, is complex and therefore tricky to use.

Many units incorporating microprocessors which may be programmed by the customer to seek for particular sequences, eg poll messages. Relatively simple devices are however available which simply show the characters being transmitted on a VDU screen. Such devices cost in the region of £1,000 while the more sophisticated devices can cost up to £10,000.

While a protocol test set can be extremely useful it is expensive and many organisations feel that the cost cannot be justified. It is only really necessary if some new untried piece of equipment or software is to be used on the network, such as a new emulation of a standard protocol.

Network Control Equipment

Network control equipment is designed to supply a wide range of diagnostic facilities which will allow fault-finding and the reconfiguration of a network under the control of a central site. This ability to monitor and test from a central location can drastically reduce the time it takes to locate a fault and permits a significant amount of diagnostic work to be carried out while the network is operational.

Network control equipment is based on the test equipment described earlier in the chapter and tests are carried out in much the same way.

Network control systems of a wide range of complexity can be obtained, the more sophisticated being designed to be built into a network when it is first implemented while simpler systems can be added at a later date. A common element in a network control system is an operator's console at the central site. Because of the high cost of a network control system it is generally found only on large networks.

10 The Final Selection Process

INTRODUCTION

Once the weighted ranking evaluation method has been applied it is then possible to make the final selection decision both for individual items and for the entire network. The decisions must be made bearing three things in mind:

- the results of the evaluation method;
- factors which relate to each individual item;
- factors which relate to the overall selection.

The results of the evaluation method, especially where this has produced no clear-cut 'winner', should be modified by other factors. It is not sufficient to select the best model for each individual item, the system must be considered as an entity before making the final decisions. It is not the purpose of this book to go into depth on this subject but the important factors are highlighted below.

SELECTION OF A SINGLE ITEM

In selecting a single item of hardware it is possible that the evaluation process will have produced scores which are similar for several models considered, none of them being clearly the best. In this case the data processing manager must decide how he will finally select between them. Provided each model measures up to the technical specification, this is usually done on cost or supplier factors such as proven design or the company's reputation. Alternatively it is possible to select the model that best satisfies the criterion with the highest weighting. The ultimate choice is dependent on the situation and no general recommendations can be made.

Regardless of whether there is an outright winner, two factors which have indirectly been treated in the weighted ranking evaluation must be further considered, namely the overall cost and the use of the model in future network development.

Overall Cost

Various elements of cost were considered as criteria in the qualitative evaluation but overall cost was not. This is because of the difficulty of directly comparing the cost of equipment from different manufacturers. Features may appear as standard in one piece of equipment, or as an extra in another; one manufacturer may only rent his equipment while another may only sell, etc. However, before making a final decision some table of comparisons of the overall cost must be drawn up. This should show clearly what each manufacturer is offering and what this will cost to your organisation over the life of the equipment, say 3 to 5 years. While it may then still not be possible to do any exact quantitative comparison, it will give a better feel for the overall picture and the best options to take.

Future Use

The average life of an item of data communications equipment is 3 to 5 years. A network is likely to develop during that period, and it is best to select equipment that will be useful throughout its life. This can be almost as important with rented equipment as with purchased, especially if the rental equipment that has the ability to be upgraded without considerable expense and without removal from service for a lengthy period.

It can also increase the value of an optional feature on a device which is not required initially, as it may well prove useful at a later date, and mean a new device does not have to be acquired.

SELECTION OF A RANGE OF EQUIPMENT

Having evaluated each item in a range of equipment, the user must now look at the configuration as a whole, firstly using the best individual items. The user should evaluate:

- the total cost of the configuration and compare this with the total budget available;
- the ability of the configuration to satisfy all the overall requirements;
- any problems posed; for example, the configuration may be operationally complex or may be too restrictive in the long term.

If there are any problems in these areas then an interative process may be necessary by selecting items that achieved lower scores in the weighted ranking evaluation until a suitable configuration is arrived at.

There are two other primary factors to be considered in the final selection of a range of equipment; namely, compatibility and the number of suppliers.

Compatibility

It is essential to ensure that all the equipment is compatible; that, for example, the multiplexer from supplier A can function correctly with the modems from supplier B. There are several ways to set about this; firstly, by ensuring that each individual piece of hardware is compatible with international standards such as CCITT standards. Secondly, the opinion of each supplier should be asked and the technical specifications of the equipment checked; and thirdly. customers that have been using a similar combination of hardware should be visited.

Number of Suppliers

Both installation and maintenance are simplified if the total number of suppliers is kept to a minimum. Some suppliers are strong in all types of data communications hardware and it is highly likely that modems, multiplexers, concentrators, racks and test equipment from one or all of them will have been considered in the selection process. Before making the final selection decisions, draw up a list showing how well each supplier's models have fared in the evaluation exercises. If, for instance, in all items except 9600 bps modems, supplier A's equipment came top of the list and its 9600 bps modems came a close second or third, it may be worthwhile considering selecting these modems as well.

It may also be more economical in that if a large volume is purchased from one supplier it may be possible to negotiate a discount.

11 A Case Study

INTRODUCTION

Consider a retail organisation with a large mainframe at its headquarters in Manchester. The mainframe is used exclusively for batch processing but there is considerable spare capacity and the organisation is considering adding new applications. The most important of these is a computerised order-entry and stock-control system which will require data communications facilities between the company's headquarters, its large warehousing sites in London, Glasgow, Bristol and Liverpool, and approximately 100 smaller retail outlets scattered throughout the UK. In addition there have been requests from the warehousing sites at Glasgow, Bristol and Liverpool for the provision of an on-line bureau service.

A five-year plan is drawn up to implement an on-line data network in three phases to provide facilities for the order-entry and stock-control system, the on-line bureau service, and to allow on-line data entry and verification for various existing applications such as payroll. The selection exercise described in the following sections concentrates on the hardware necessary; Post Office services and terminals, and network design, are not discussed in depth. The aim is to show the evaluation exercise in use in a practical situation.

A FIVE-YEAR PLAN

Development of Use of the Network

Phase 1

This is shown logically in Figure 11.1. All four warehousing sites will be allowed access to the computer in Manchester during Phase 1 in such a way that Phase 2 and 3 can build on this original configuration.

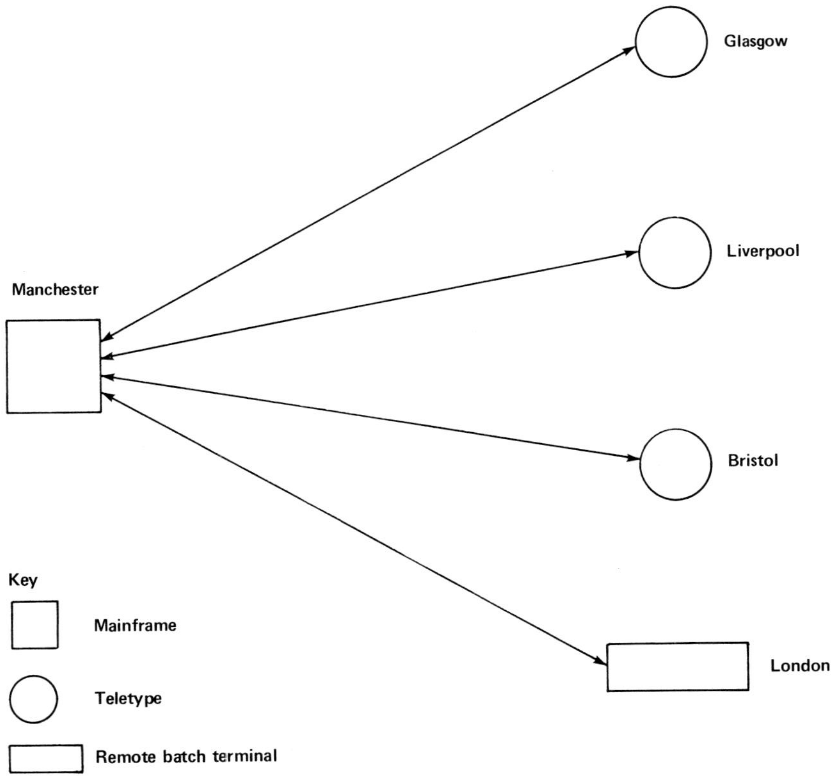

Figure 11.1 Logical arrangement of Phase I

London	– A remote job entry terminal (RJE) is installed to allow computerisation of existing manual systems in connection with the order-entry system. In addition, other manual systems in operation may be computerised and batch jobs submitted.
Glasgow Bristol Liverpool	– 1 teletype is required at each site to allow access to the computer for program development and for the use of a scheduling software package. The volume of use of these teletypes is expected to be in the region of one hour each per working day.

When the RJE system installed in London is working satisfactorily a similar arrangement will be installed in Glasgow. The warehouses at Liverpool and Bristol are much smaller and are not likely to require such an installation.

Phase 2

During Phase 2, the logical arrangement of which is shown in Figure 11.2, the installation at London is to remain unchanged. Similarly no extra equipment is required at Bristol or Liverpool although volume of usage of the teletypes is planned to grow as the terminals are to be used for on-line data entry as well as the bureau service.

At Glasgow the teletype is to be used for on-line data entry as well as the bureau service. In addition however Scotland is to be first area where VDUs are to be installed in retail outlets. One teletype compatible VDU will be installed at the warehouse site in Glasgow and one in each of 13 retail outlets.

These will be used for entering stock levels and orders directly to the computer at Manchester. It is estimated that they will be used at random times during the working day but total usage will be low – in the region of 1 hour a day.

Phase 3

The logical arrangement for Phase 3 is shown in Figure 11.3. The arrangement in Scotland is expanded so that VDUs are installed in all retail outlets there, while the Bristol and Liverpool areas have VDUs installed for the first time. 10 VDUs are also installed in retail outlets in the Manchester area. The London installation remains unchanged because the working arrangements are different in the area, orders being placed directly with the warehouse and entered to the computer at Manchester via the RJE system. The volume of work has increased sharply however. An RJE system

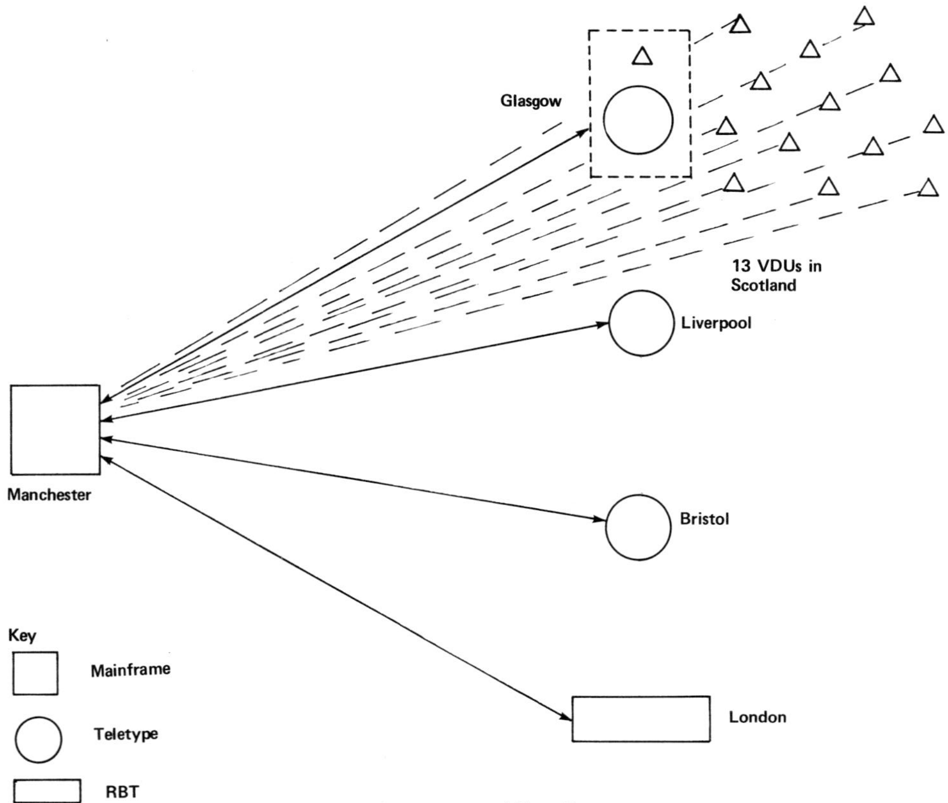

Figure 11.2 Logical arrangement of Phase II

is also now installed in Glasgow to deal with manual systems which have been computerised, as in London.

Bristol – One asynchronous teletype is required at the warehouse site for access to the bureau service and for on-line data entry. One VDU is required for placing of orders, in addition a VDU is placed in 28 retail outlets in Wales and the South West. These will be used for entering orders to the computer at Manchester at random times during the day. Total use of each VDU will be less than one hour a day.

Liverpool – One asynchronous teletype is required at the warehouse site for access to the bureau service and for on-line data entry and one VDU for use of the order-entry system. In addition a VDU is placed in 10 retail outlets in the area. These will be used for entering orders to the computer at Manchester at random times during the day. Total use of each VDU will be less than one hour a day.

Glasgow – One remote batch terminal at the warehouse site and one VDU for use with the order-entry system. In addition a VDU is placed at 36 additional retail outlets in Scotland, making a total of 51 remote VDUs. Use of each VDU will be less than one hour a day.

Additional Information on Use of the Network

Certain general information is necessary before the network configuration can be drawn up for any of the three phases:

- any asynchronous teletype selected will operate at 30 characters per second and half-duplex transmission is to be used;
- the asynchronous teletype to be installed in Glasgow during Phase 1 will be in a fixed position while those in Liverpool and Bristol will be shared by several departments and must hence be portable;
- any remote batch terminal selected will operate synchronously and full-duplex transmission is to be used;
- any leased line installed must have PSTN back up facilities;
- the number of suppliers is to be kept to a minimum;
- all VDUs are teletype-compatible and we assume they operate asynchronously at 30 characters per second.

Configuration of the Network

From the logical network diagrams, shown in Figures 11.1, 11.2 and 11.3, and the information in the preceding sections of this chapter the physical configuration of the network during the three phases of its development can be constructed.

Phase 1

Figure 11.4 shows the network configuration that can be constructed for Phase 1. A 300 bps PSTN circuit will be adequate for the links Glasgow to Manchester, Liverpool to Manchester and Bristol to Manchester. The London to Manchester link however will carry sufficient traffic to justify a leased line at 2400 bps. Some form of PSTN backup must be arranged however for this link.

Therefore, the most appropriate Post Office Datel services would be:

- Datel 200 for Glasgow to Manchester, Liverpool to Manchester and Bristol to Manchester. Modem 2s or 13s may be used with this service. The Modem 2, however, offers an autoanswer facility and is therefore preferable.
- Datel 2400 with Modem 7s which provide for PSTN fallback at 600 or 1200 bps for London to Manchester. (This requires a 4-wire circuit.) The Modem 7 allows for autoanswering when operating on the PSTN but has no error detection or correction facilities.

However, because of the need for portability at Liverpool and Bristol, an acoustic coupler would be more appropriate than a Modem 2 at each site. An acoustic coupler is able to work satisfactorily with a Modem 2.

At Manchester some device will be necessary to connect the data communications links to the host computer. In addition, a basic set of test and diagnostic equipment as described in Chapter 9 is necessary, even though most of the equipment will be Post Office supplied.

Therefore a complete list of data communications equipment to be selected for Phase 1 is:

- 2 acoustic couplers;
- a device to connect the 4 data communications lines to the mainframe;
- a modem test set.

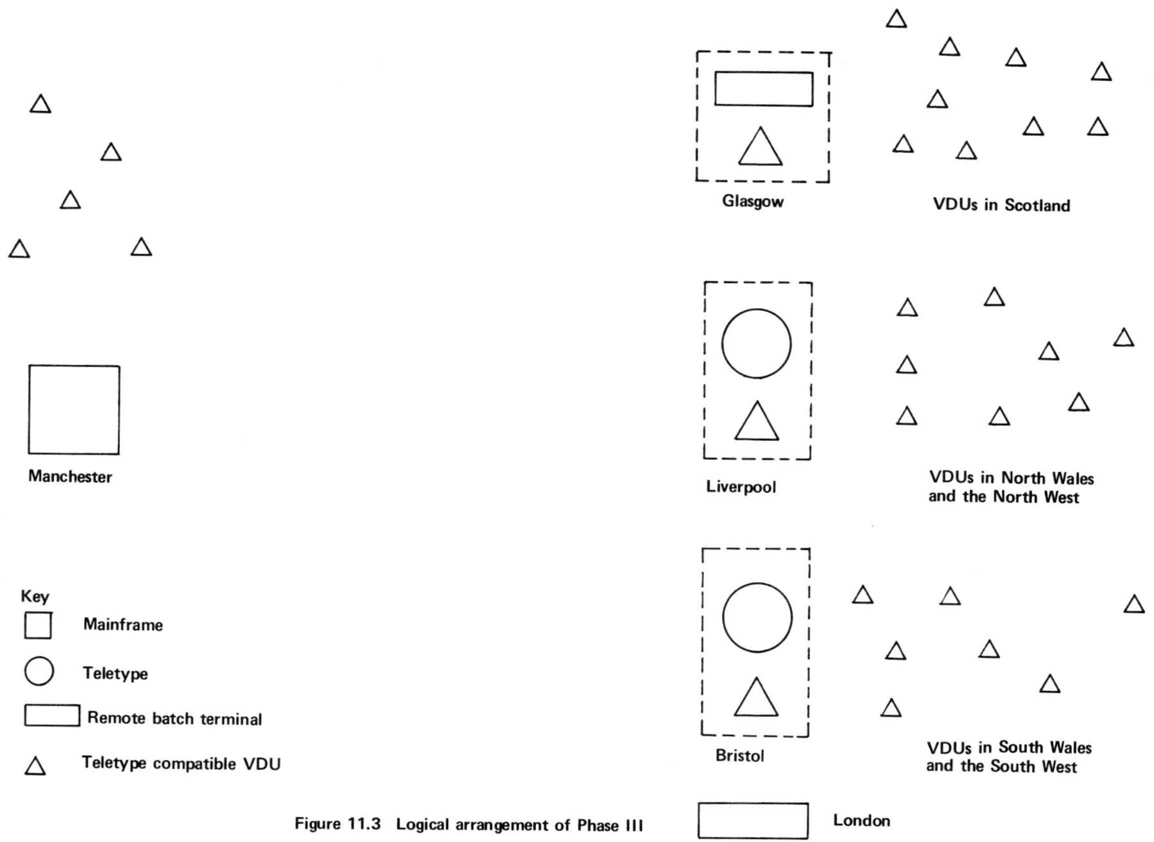

Figure 11.3 Logical arrangement of Phase III

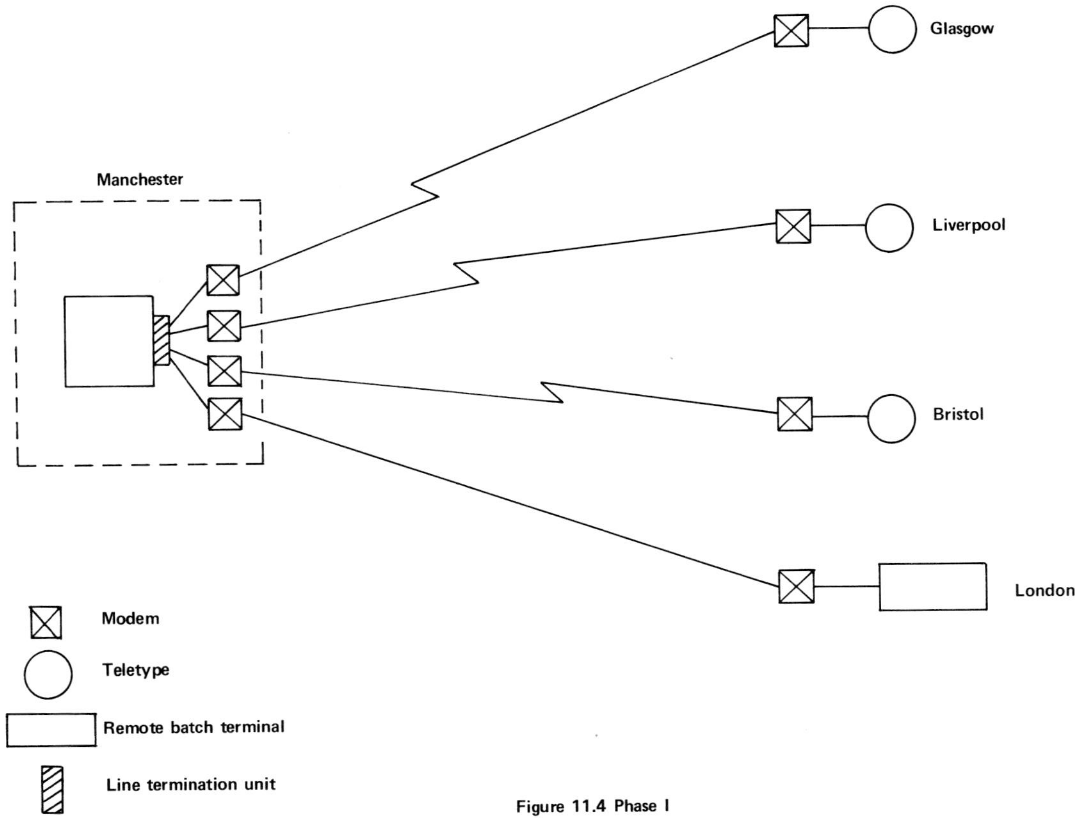

Figure 11.4 Phase I

Phase 2

Figure 11.5 shows the network configuration that can be constructed for Phase 2. Changes are only required in the Glasgow to Manchester link. Here 15 devices must be able to access Manchester from Scotland. This could be done either by use of multipoint lines or a multiplexer. All the devices could not be supported on one multipoint line; three or possibly more would be necessary for an adequate service and the long distance from Glasgow to Manchester would have to be covered and payed for more than once.

Using a multiplexer, the distance from Glasgow to Manchester would only have to be covered by one link, albeit at a higher speed. In Phase 3 more devices will be added in Scotland however and a multiplexer or concentrator will then almost certainly be necessary. It would therefore be more sensible to establish the initial configuration during Phase 2 than have to dismantle a multipoint configuration and begin again in Phase 3.

The teletype-compatible VDU and teletype at the warehouse site may be connected directly to the multiplexer. The remote VDUs must be connected via a Datel service. As use of the VDUs is low a dial-up link from each to the multiplexer would be the most economical method. The most appropriate Datel service would be Datel 200 with Modem 2C.

Using a multiplexer all 15 devices, each capable of operating at 30 characters per second, may gain access to the computer simultaneously. Therefore the line between Glasgow and Manchester must have sufficient capacity for this, the nearest speed available from the Post Office being 4800 bps. This will be a leased line because with 15 devices sufficient traffic is generated to make use of this line fairly high. Also at 4800 bps the error rate on the PSTN is unacceptably high. At 4800 bps the Post Office does not offer a Datel service, the line may be leased but proprietary modems with PO approval must be used.

Therefore a complete list of data communications equipment to be selected for Phase 2 is:

- two multiplexers;
- two 4800 bps modems.

Phase 3

Figure 11.6 shows the network configuration that can be constructed for Phase 3. Because of the increased traffic on the

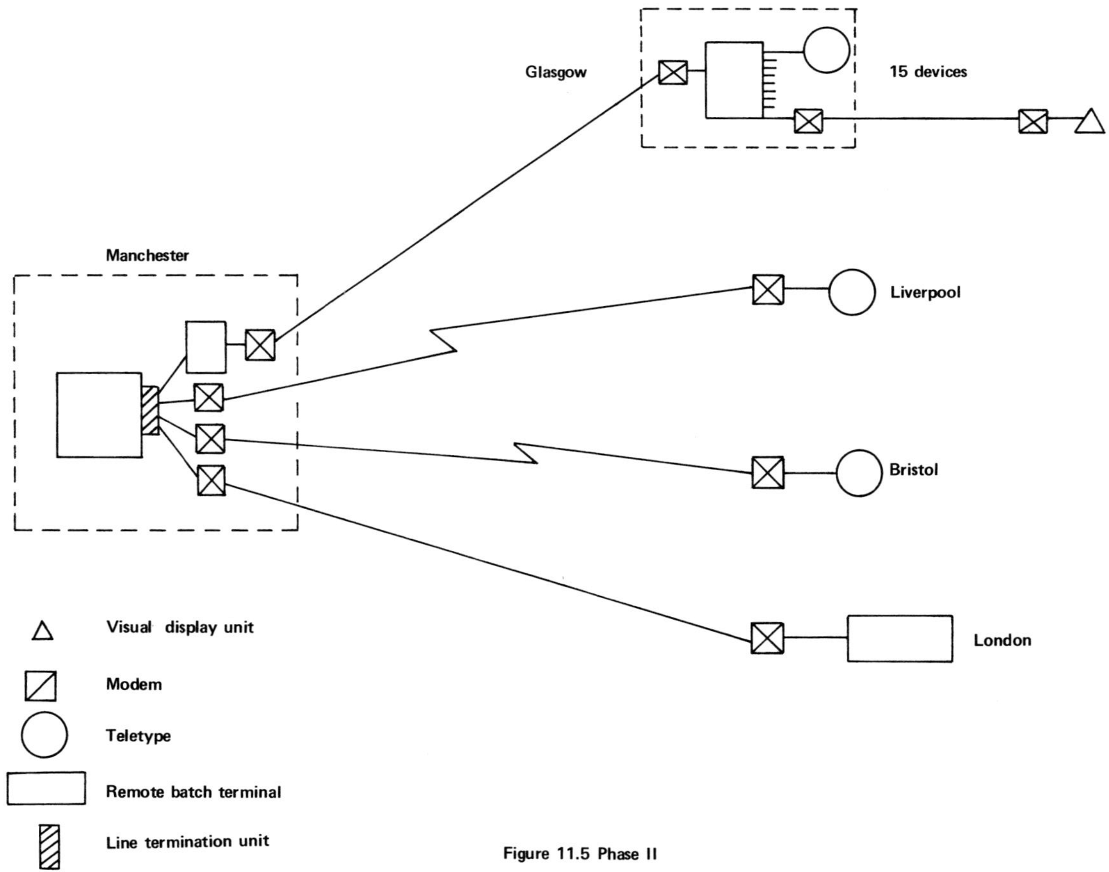

Figure 11.5 Phase II

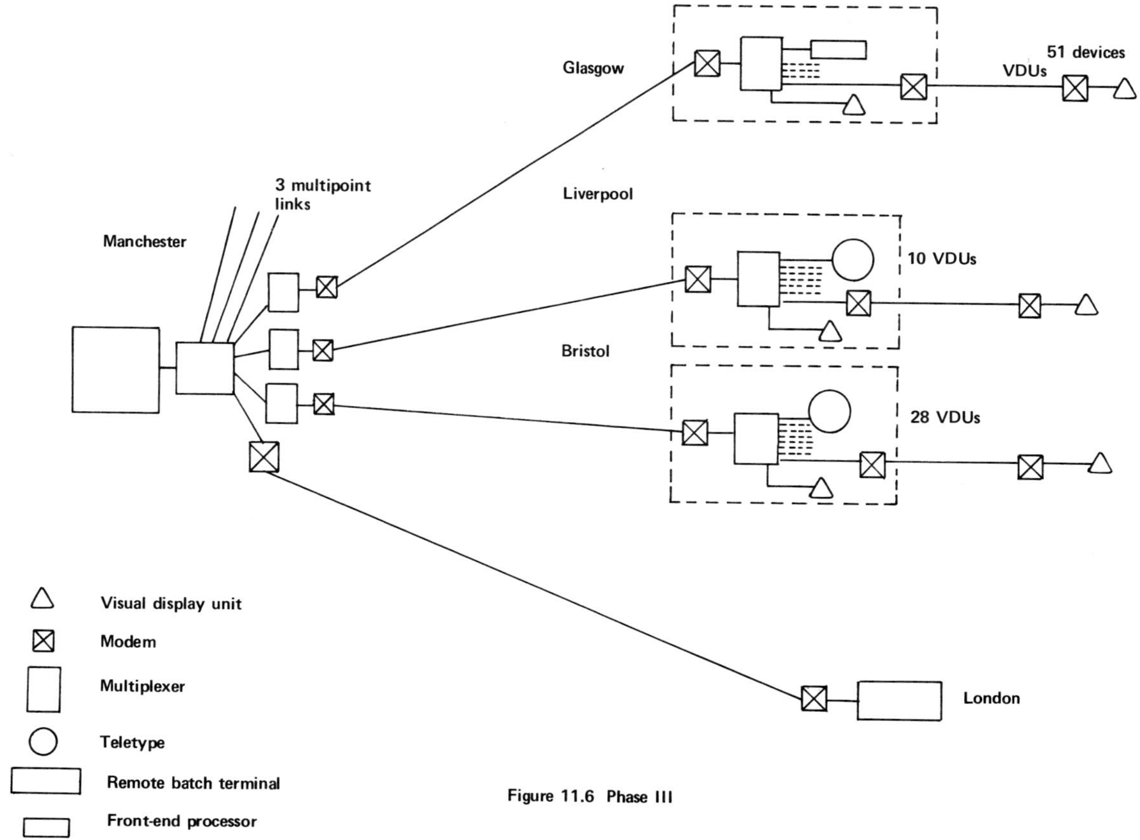

Figure 11.6 Phase III

London to Manchester link it must be upgraded to 4800 bps. This requires the use of a proprietary modem.

The Bristol and Liverpool configurations must be modified to resemble the arrangement implemented at Glasgow during Phase 2. The number of devices however is different in each case, 30 in Bristol and 12 in Liverpool. This results in the need for a 9600 bps Post Office leased line between Bristol and Manchester and a 4800 bps line between Liverpool and Manchester. Proprietary modems are therefore required on each link.

The arrangements at Glasgow are modified once again during this phase. The asynchronous teletype is replaced by a remote batch terminal, similar to that installed in London and 37 more retail outlets are fitted with VDUs. This gives a total of 51 asynchronous devices and 1 synchronous device which must be able to access the Manchester computer. The existing multiplexer could not cope with this traffic and must therefore be replaced by a device, either an intelligent multiplexer or concentrator, which employs contention and can deal with several output links to Manchester. Alternatively to reduce the vulnerability of the system to multiplexer failure, several simple multiplexers could be used.

In addition, 10 VDUs in the Manchester area are connected by 3 multipoint links at 1200 bps to the mainframe. The system has now expanded considerably, supporting VDUs, two remote batch terminals and two teletypes. The mainframe is in need of help with the communications processing to prevent overload and consequently a front-end processor is required. Therefore the total data communications hardware to be selected for Phase 3 is:

- several 4800 and 9600 bps modems;
- an intelligent multiplexer or concentrator;
- a front-end processor.

EQUIPMENT SELECTION FOR PHASE 1

To recapitulate, the non-Post Office equipment to be selected for Phase 1 is:

- 2 acoustic couplers;
- a device to connect 4 data communications lines to the mainframe;
- a modem test set.

Selection of the first two items is relatively straightforward and does not require the application of the weighted ranking method. Each item is discussed individually below.

The Acoustic Couplers

The acoustic couplers selected may either be independent devices or form part of the terminal. The selection of the acoustic couplers must therefore be made in conjunction with the terminal equipment. Terminals which incorporate an acoustic coupler tend to be of the portable variety. The difference between various models of acoustic coupler is minimal, hence there is no need for a detailed evaluation of each model. Factors which must be borne in mind however arise from the general criteria discussed in Chapter 4. Specifically these are:

- is the acoustic coupler compatible with a modem?
- do you wish to rent or buy the device? The average purchase price for an acoustic coupler is £150 to £200 while the monthly rental is in the region of £10 to £15 depending on the length of the rental contract.

Mainframe Interface Device

A device is necessary to interface the four data communications lines to the mainframe at Manchester. This device must be procurred from the mainframe manufacturer and no evaluation exercise is therefore necessary. The device must be able to support the required line speeds, ie 300 and 2400 bps, the required line protocols and, of course, the necessary number of lines. It is unlikely that the manufacturer will offer a range of devices which satisfy these requirements so selection will be straightforward.

Modem Test Set

This item forms part of the required basic test equipment. The simple configuration of Phase 1, with its almost exclusive use of Post Office services, does not make its acquisition urgent. However the organisation has decided that it wishes to keep network availability high and that experience gained in fault-finding during the early stages will be invaluable in later phases. It has therefore decided to buy a modem test set during Phase 1. In order that the test set will be useful during Phase 2 and 3 of the networks development it must be flexible. Bearing in mind both technical and general criteria the following are a list of essential criteria:

– operates at 300–4800 bps	ie matches
– operates asynchronously and synchronously;	modem
– operates half-duplex and duplex;	features

– V24 interface;

– pseudorandom pattern transmission;
– interface monitoring;
– portability;
– Post Office approved.

The desirable criteria are:

– reliable, in terms of MTBF;
– maintained by supplier;
– supplier credibility;
– proven design;
– product range to be supported/expanded in future;
– basic cost;
– maintenance cost;
– bought or rented?
– block error count.

Three test sets are considered, as shown in Table 11.1. Each of these satisfies all the essential criteria.

The scoring schemes used for the desirable criteria are as follows:

– Block error count	10 points if present 1 point if not
– Bought or rented	10 points if rental available 1 point if not
– all others	10 for best model and lesser scores for others on a sliding scale.

EQUIPMENT SELECTION FOR PHASE 2

To recapitulate, the non-Post Office equipment to be selected for Phase 2 is:

– two multiplexers;
– two 4800 bps modems.

	Reliability	Maintenance	Supplier credibility	Proven design	Product ramge	Basic cost	Maintenance cost	Bought or rent	Block error
Reliability (MBTR	–	1	1	2	1	2	2	1	2
Maintenance	3	–	1	2	1	1	2	1	2
Supplier credibility	3	3	–	3	2	3	3	1	3
Proven design	2	2	1	–	1	2	3	1	2
Product range	3	3	2	3	–	3	2	2	3
Basic cost	2	3	1	2	1	–	1	2	2
Maintenance cost	2	2	1	1	2	3	–	2	3
Bought or rented	3	3	3	3	2	2	2	–	3
Block error	2	2	1	2	1	2	1	1	–
Totals	20	19	11	18	11	18	16	11	20
Weightings	8	7	2	6	2	6	5	2	8

Criterion	Weighting	Model A		Model B		Model C	
		Score	Weighted score	Score	Weighted score	Score	Weighted score
Reliability	8	7	56	9	72	6	48
Maintenance	7	6	42	7	49	6	42
Supplier credibility	2	6	12	6	12	10	20
	6	8	48	8	48	8	48
Product range	2	5	10	5	10	6	12
Basic cost	6	9	54	8	48	5	30
Maintenance cost	5	8	40	9	35	5	25
Bought or rented	2	10	20	10	20	1	10
Block errors	8	10	80	10	80	1	8
Totals			362		374		223

Table 11.1 Calculations of weightings and weighted scores for modem in Phase 1

Multiplexers

Two simple multiplexers are required, one at Glasgow and one at Manchester. The initial decision must be whether these are to be based on frequency division or time division multiplexing techniques. As the multiplexer must be able to deal with 15 devices all capable of operation at 30 characters per second a frequency division multiplexer would not be adequate. A character TDM multiplexer would be the most suitable as all the devices which will be attached to the multiplexer are asynchronous. The essential criteria are identified as:

- ability to handle 15 asynchronous devices as input;
- input speeds of 300 bps;
- EBCDIC code;
- one output channel of 4800 bps;
- delivery within 13 weeks of order being placed;
- rack-mounted;
- Post Office approval.

The desirable criteria are identified as:

- basic cost;
- maintenance cost;
- reliability in terms of mean time between failure;
- level of maintenance;
- supplier credibility;
- good monitoring and testing facilities;
- simplicity of installation;
- proven design;
- any existing hardware from this supplier;
- product range is to be maintained/expanded?

The evaluation exercise is carried out in Table 11.2. All the desirable criteria are scored on a sliding scale from 10 to 1 except the criterion 'any existing hardware from supplier'. In this case 10 is awarded for a positive response and 1 for a negative.

Modems

Two 4800 bps modems are required, one at Glasgow and one at

	Basic cost	Maintenance cost	Reliability (MTBF)	Maintenance	Supplier credibility	Monitoring & diagnostic	Simple installation	Proven design	Hardware already	Product range
Basic cost	–	1	2	2	1	2	1	2	3	2
Maintenance cost	3	–	2	2	1	3	1	2	3	2
Reliability	2	2	–	1	1	2	1	2	2	1
Maintenance	2	2	3	–	2	2	1	2	3	1
Supplier credibility	3	3	3	2	–	3	2	2	3	1
Monitoring & diagnostic	2	1	2	2	1	–	1	2	2	1
Simple installation	3	3	3	3	2	3	–	2	3	2
Proven design	2	2	2	2	2	2	2	–	3	2
Hardware already	1	1	2	1	1	2	1	1	–	3
Product range	2	2	3	3	3	3	2	2	3	–
Totals	20	17	22	18	14	22	12	17	25	15
Weightings	6	5	7	5	3	7	2	5	9	4

Criterion	Weighting	Model A		Model B		Model C	
		Score	Weighted score	Score	Weighted score	Score	Weighted score
Basic cost	6	10	60	8	48	8	48
Maintenance cost	5	6	30	6	30	7	35
Reliability (MTBF)	7	7	49	4	28	9	63
Maintenance	5	7	35	5	25	7	35
Supplier credibility	3	8	24	4	12	8	24
Monitoring & diagnostic	7	8	56	4	28	7	49
Simple installation	2	7	14	5	10	5	10
Proven design	5	6	30	5	25	9	45
Hardware already	9	1	9	1	9	1	9
Product range	4	8	32	4	16	7	28
Totals			339		231		346

Table 11.2 Calculations of weightings and weighted scores for multiplexer in Phase 11

Manchester for the link between the simple multiplexers. The Post Office does not at present supply such a device so a proprietary modem is required. The essential criteria are identified as:

- modem operates at 4800 bps, synchronously on a leased line;
- modem operates in full-duplex mode;
- Post Office approved;
- rack-mounted;
- compatible with multiplexer, V24 interface;
- delivery within 13 weeks of order being placed;
- fallback facility to PSTN at 1200 bps available.

The desirable criteria are identified as:

- type of equalisation;
- capable of operation at 9600 bps (or upgradable to 9600 bps);
- extensive diagnostic facilities;
- voice adaptor facility;
- short turn-around time;
- easy to operate;
- reliability (MTBF);
- basic cost;
- simplicity of installation;
- maintenance support;
- maintenance cost;
- is the other equipment from that supplier already in the network?

Three models are identified which satisfy all the essential criteria and these are evaluated using the weighted ranking method in Table 11.3. Because there will be a requirement for 9600 bps modems during Phase 3 it may be useful if the modem is upgradable to operate at this speed.

EQUIPMENT SELECTION FOR PHASE 3

To recapitulate, the non-Post Office equipment to be selected for Phase 3 is:

	Equalisation	Operation at 9600 bps	Diagnostic facilities	Voice adaptor facility	Short turnaround time	Ease of operation	Basic cost	Reliability	Simplicity of installation	Maintenance support	Maintenance cost	Supplier already
Equalisation	–	1	2	3	2	1	1	2	1	1	1	2
Operation at 9600 bps	3	–	2	3	3	1	2	3	1	3	1	3
Diagnostic facilities	2	2	–	2	2	1	2	2	1	3	2	3
Voice adaptor	1	1	2	–	2	1	2	3	1	1	1	2
Short turnaround time	2	1	2	2	–	1	2	3	1	2	2	2
Ease of operation	3	3	3	3	3	–	3	3	2	3	3	3
Basic cost	3	2	2	2	2	1	–	2	1	2	1	2
Reliability (MTBF)	2	1	2	1	1	1	2	–	1	2	1	2
Simplicity of installation	3	3	3	3	3	2	3	3	–	3	3	3
Maintenance support	3	1	1	3	2	1	2	2	1	–	2	2
Maintenance cost	3	3	2	3	2	1	3	3	1	2	–	3
Supplier already	2	1	1	2	2	1	2	2	1	2	1	–
Totals	27	19	22	27	24	12	24	28	12	24	18	27
Weighting	7	4	5	7	6	1	6	8	1	6	3	7

Criteria	Weighting	Model A		Model B		Model C	
		Score	Weighted score	Score	Weighted score	Score	Weighted score
Equalisation	7	10	70	10	70	10	70
Operation at 9600 bps	4	10	40	1	4	10	40
Diagnostic facilities	5	8	40	8	40	6	30
Voice adaptor	7	10	70	10	70	1	7
Short turnaround time	6	8	48	6	36	4	24
Ease of operation	1	8	8	6	6	6	6
Basic cost	6	6	36	8	48	8	48
Reliability (MTBF)	8	9	72	6	48	10	80
Simplicity of installation	1	5	5	8	8	5	5
Maintenance support	6	8	48	5	30	8	48
Maintenance cost	3	6	18	5	15	8	24
Supplier already	7	10	70	1	7	1	7
Totals			525		378		389

Table 11.3 Calculations of weightings and weighted scores for modem in Phase II

- a multiplexer or concentrator;
- a front-end processor;
- several 4800 bps and 9600 bps modems.

A Multiplexer or Concentrator

Because of the expansion of the terminal system in Scotland, additional multiplexing or concentrating facilities are required in Glasgow. There are three solutions:

- use of several TDM multiplexers;
- use of an intelligent multiplexer;
- use of a concentrator.

The simplest solution is to use two simple multiplexers with 9600 bps links for the asynchronous devices and a separate Datel service with a 2400 bps link for the RBT. Assuming the distance from Glasgow to Manchester is 200 miles, that a 9600 bps modem costs approximately £3,500, and a suitable multiplexer £10,500, the total cost of this solution is in the region of £450 rental per month plus an initial outlay of £56,500 (assuming multiplexers and modems are purchased).

Using either an intelligent multiplexer or concentrator, all the devices, including the RBT, can make use of one 9600 bps link. The monthly rental to the Post Office for such a system would be £200 and an initial outlay of £7,200 for modems and PO connection. Therefore provided any intelligent multiplexers or concentrators purchased cost less than £4,900 this solution would be more economic. However the use of contention would increase the response time to users. In addition, all the terminals will be dependent on one link and one device; if either fails all terminals lose access to the computer centre. In this example, the organisation decided to install an intelligent multiplexer (because individual terminal usage is low) with a spare modem at Glasgow and a fallback link for the 9600 bps link.

The essential criteria for the intelligent multiplexer are identified as:

- delivery within 26 weeks of order being placed;
- rack-mounted;
- able to handle 51 asynchronous and 1 synchronous input;
- able to handle 300 bps and 2400 bps inputs;

- 2 synchronous 9600 bps output (one for fallback);
- EBCDIC code;
- Post Office approved;
- Compatible with modems and front-end processor.

The desirable criteria are identified as:

- simplicity of installation;
- basic cost;
- maintenance service;
- maintenance cost;
- error detection and correction facility;
- good-test and diagnostic features;
- supplier credibility;
- does supplier already supply hardware for the network?
- is product range to be developed/extended?

Two multiplexers initially considered satisifed all the essential criteria and these are evaluated in Table 11.4. The simple multiplexer used at Glasgow during Phase 2 may be transported to Bristol or Liverpool for use there during Phase 3.

A Front-End Processor

During Phase 3 there are 7 telecommunications lines connected to the host computer at Manchester, allowing 103 terminals access to the system. This is above the load that simple line adaptor units can efficiently deal with and a front-end processor is required. It is decided that a passive FEP obtained from the mainframe manufacturer is the most appropriate device. Having made this decision, the selection process is reduced to drawing up a list of criteria as the specification for the manufacturer's FEP. These criteria are:

- no modifications necessary to existing hardware;
- minimal disruption to existing software;
- the manufacturer will supply on a turnkey basis;
- 84 half-duplex and 2 duplex lines can be handled simultaneously;
- handles the correct line protocols;

	Simplicity of installation	Basic cost	Maintenance service	Maintenance cost	Error detection and correction	Good test and diag. facilities	Supplier credibility	Supplier already	Product range
Simplicity of installation	–	2	2	1	2	2	1	2	1
Basic cost	2	–	2	1	3	2	1	2	1
Maintenance service	2	2	–	2	2	2	1	2	1
Maintenance cost	3	3	2	–	3	3	2	3	2
Error detection and correction	2	1	2	1	–	2	1	2	1
Good test and diagnostic facilities	2	2	2	1	2	–	1	2	1
Supplier credibility	3	3	3	2	3	3	–	2	1
Supplier already	2	2	2	1	2	2	2	–	1
Product range	3	3	3	2	3	3	3	3	–
Totals	19	18	18	11	20	19	12	18	9
Weighting	7	6	6	2	8	7	3	6	1

Criteria	Weighting	Model A		Model B	
		Score	Weighted score	Score	Weighted score
Simplicity of installation	7	5	35	5	35
Basic cost	6	9	72	6	36
Maintenance service	6	8	48	8	48
Maintenance cost	2	9	18	5	10
Error detection and correction	8	9	72	8	64
Good test and diagnostic facilities	7	6	42	8	56
Supplier credibility	3	8	24	8	24
Supplier already	6	1	6	10	60
Product range	1	8	8	8	8
Totals			325		341

Table 11.4 Calculations of weightings and weighted scores for multiplexers in Phase III

– connects to one host.

4800 and 9600 bps Modems

During Phase 3, the modems required are:

Liverpool to Manchester	4800 bps modems
Bristol to Manchester	9600 bps modems
Glasgow to Manchester	9600 bps modems
Multipoint lines at Manchester	Post Office modems

As the organisation already has acquired proprietary 4800 bps modems during Phase 2 then further modems from this manufacturer should certainly be considered and selected unless they cannot provide any special feature which is required. It is likely that if a manufacturer's 4800 bps modems operate satisfactorily, the 9600 bps will be similar and an effort should be made to keep the number of suppliers to a minimum.

The evaluation exercise will not be detailed here, as an example of a modem evaluation has already been carried out.

12 Conclusions

Today there is a wide range of data communications hardware available. This, coupled with the high cost of such equipment, makes a methodical approach essential in any selection exercise. The weighted ranking method has proved an effective tool in this area. The basic steps in this evaluation method are identical for all types of hardware, these steps are:

– draw up a list of criteria on which evaluation is to be based;

– identify essential criteria;

– award weightings to remaining criteria;

– award scores to each criteria;

– evaluate each model;

– compare results and make selection.

The criteria can be divided into two groups, which we have called general and specific criteria. The general criteria are those which are equally applicable to all items of data communications hardware; the specific criteria are different for each type of hardware as they relate to its technical features. The general criteria were discussed in Chapter 4 while the specific criteria are discussed in Chapter 5 to 9, where the different types of data communications hardware are described.

A complete list of criteria has been drawn up for each type of equipment but it is up to the individual to decide the relevance of each of these to a particular situation. Some criteria will always be essential for instance compatibility, while others may be essential in one set of circumstances and totally irrelevant in another. Before considering different models of equipment, the relative importance of each criterion must be decided.

The scoring schemes to be used for each criterion may also vary from one selection exercise to another. Scoring schemes have been suggested for both general and specific criteria but the data processing manager may find that others are more appropriate to his particular situation.

The evaluation method, which is discussed in detail is in itself not sufficient for the selection of the most appropriate pieces of hardware. Its rigorous approach must be tempered by overall considerations such as cost and the requirements of the overall configuration.

Network design has been discussed because the hardware selection cannot be carried out in isolation. It is often some requirement identified in the design stage that is the deciding factor in the selection of an item of equipment. If, for example, only one item is being selected to add to an existing network then the overriding factor is compatibility with the existing equipment. Similarly the technical features of an item are generally defined by the overall requirements of the network. Although only an outline of the considerations of network design is given in Chapter 3 those factors which influence the equipment selection are emphasised.

The case study in Chapter 11 shows not only a practical application of the selection process, but the design considerations which lead to the requirement for particular data communications hardware. The case study is a hypothetical example used to highlight particular aspects of the evaluation method.

Appendix 1 Glossary

Amplifier	– Device which increases the power of an electric signal
Analogue Transmission	– Transmission of a continuously variable signal (usual transmission method for telephone lines) as opposed to a discretely variable signal (as in digital transmission)
Attenuation	– Decrease in magnitude of current voltage or power of a signal in transmission
Bandwidth	– Range of frequencies available for signalling in a communications channel. Usually expressed as the difference between the highest and lowest limits of the frequency range which can be used (eg bandwidth of a voice grade channel = 2700 Hz which is the difference between the lowest unit of 300 Hz and the highest limit of 3000 Hz).
Baud	– Unit of discrete signalling speed per second, the modulation rate. Bauds = bits per second when two state signalling is used
Bit	– Abbreviation of 'binary digit'. Signal element of transmission in binary notation either '0' ('off') or '1' ('on')
Bps	– Bits per second Commonly used units for expressing data signalling rate

Broadband – A communications channel having a bandwidth greater than a voice grade telephone circuit. Also called 'wide-band'

Byte – A group of consecutive bits operated on a unit by a computer, commonly 8 bits

CCITT – Comite Consultatif International Telegraphique et Telephonique (International Consultative Committee for Telegraphy and Telephony)

Character – Letter, figure, number, punctuation or other sign contained in a message; usually represented by one byte

Circuit – Means of two-way communication between two points comprising a 'send' and 'receive' channel.

Common Carrier – National organisation which has the authority to provide public telecommunications services in its own country.

Conditioning – Improvement of ordinary telephone lines for data transmission by means of special equipment to carry out equalisation etc

DATEL – The Post Office data transmission services comprising a suitable line, modem and, where necessary, maintenance support

DCE – Data Circuit Terminating Equipment equipment installed which establishes, maintains and terminates the connection and carries out signal conversion and coding on the line (eg a modem or equivalent)

Dial-Up – Means of obtaining a communication link on the Public Switched Network by dialling a telephone number

Digital Transmission	– Transmission of a discretely variable signal, a signal whose various states are discrete intervals apart
Distortion	– Undesirable change in waveform which occurs between two points in a transmission system
DTE	– Data Terminal Equipment User's equipment which is attached to the network to send and/or receive data
Echo	– A signal which has been partly reflected or otherwise returned with sufficient magnitude and delay for it to be perceptible in the circuit after the original signal has been transmitted
Echoplex	– Return by the receiver of a character from an input keyboard to the local output printer
Echo Suppresor	– Device which reduces echoing on the circuit; it attentuates the transmission path in one direction while allowing signals to pass in the other direction
Error Rate	– Ratio of number of bits, elements characters or blocks incorrectly received to the total number of bits, elements, characters or blocks sent
Filter	– A device which transmits signals of frequencies within one or more specified frequency bands and attentuates signals of other frequencies
HDLC	– High Level Data Link Control An international standard for frame structures in connection to data communications protocols
Hertz (Hz)	– The international standard measurement of frequency expressed in cycles per second
Kbps	– Kilobits per second

Link – Continuous data communications path between two places, eg telephone, telegraph or telex line, microwave or satellite connection

MTBF – Mean-time between failure
Measure of reliability of system; gives average period of time in which system can operate without breaking down

Multidrop – Connection of two or more terminals to a single communications link in such a way that each terminal may communicate with the central site but not with each other

Multipoint – See 'multidrop'

Parity – Condition of having an odd number of '1' bits or an even number of '1' bits in a set of data

Parity Check – Method of error control. Total number of '1' bits in a set of data is made to be either an 'odd' or 'even' number. This is done by adding an extra bit (parity bit) to make the number of '1' bits either odd or even, whatever is required

Point-To-Point – Direct link between two points in a network, eg between a computer and terminal without any intermediary device (except modems)

Port – Part of data communications device which acts as input/output connection, eg computer port

Protocol – A strict procedure required to initiate and control communication over a communications link

PSTN – Public Switched Telephone Network

PTT – Abbreviation for Postes, Tèlègraphes et Tèlècommunications, (Postal, Telegraph and Telephone authority). Name given to the telecommunications authorities

	in Europe, and also elsewhere who act as the common carriers for telecommunications (eg the Post Office in their respective countries.
SDLC	– Synchronous Data Link Control, on IBM link protocol similar to HDLC
Sideband	– Frequency band on either the upper or lower side of the carrier frequency within which fall the frequencies produced by modulation
TCU	– Transmission control unit. Also called 'communications controller' or 'communications control unit'. This device acts as an interface between the host computer and a communications network.
Voice Grade Band	– Channel which is 4000 Hz wide and capable of carrying speech signals in the range 300–3000 Hz

Appendix 2 Bibliography

Books

1 Down, P J, *Fault Diagnosis in Data Communications* (NCC Publications, 1977)

2 *Handbook of Data Communications – UK Post Office* (NCC Publications, 1975)

3 Morgan, L. *Management of On-line Data Communications Systems* (NCC Publications, 1979)

4 Bingham and Davies, *Planning for Data Communications* (MacMillan Press, 1977)

5 Pritchard, J A T, *Selection and Use of Terminals in On-Line Systems* (NCC Publications, 1978)

6 Martin, J, *Systems Analysis for Data Transmission* (Prentice Hall, 1972)

7 TARIFICA: Survey of Telecommunications Tariffs and Services in Western Europe (Logica 1976, 1977, 1978)

8 Taylor, F. E. *Teleprocessing Monitor Packages* (NCC Publications, 1979)

9 Hebditch, D, *Teleprocessing Monitor Packages for IBM 370* (NCC Publications, 1978)

10 Bleazard, G B, *Teleprocessing Monitor Packages for ICL 2903/04* (NCC Publications, 1978)

NCC COMPUTER HARDWARE RECORDS

Volume 1	Issue 6	Communications Control Hardware
Volume 1	Issue 12	Modems/Acoustic Couplers
Volume 2	Issue 8	Communications Control Hardware and Remote Batch Systems
Volume 2	Issue 12	Peripheral and Data Transmission Test Equipment
Volume 3	Issue 8	Modems/Acoustic Couplers
Volume 3	Issue 11	Small Scale Communications Controllers and Data Concentrators

Articles

1 Berglund, R G, Comparing Network Architectures, *Datamation,* February 1978

2 Hoinski, M J, *How to Specify and Evaluate Proposals for Modems,* Telecommunications, July 1975

3 Dick, G M, The Lowly Modem, *Datamation,* March 1977

4 Scharf, T, Weighted Ranking by Levels, *IAG Quarterly,* Vol 2, 2, 1969

Appendix 3 Sources of Information

THE UK POST OFFICE

The UK is divided into ten telecommunications regions and 73 telephone areas. Contact with the Post Office for a user interested in telecommunications services can be at either of these two levels or at headquarters level, although the initial point of contact should be at local area level. Each telephone area has a General Manager to whom initial enquiries should be directed; the address and telephone number will be found in your local directory. Cases are referred to the regional level if the Post Office services required are likely to involve more than one area within a region or another area in a different region. Advice from telecommunications headquarters would be sought if meeting the user's need would involve the Post Office in national policy considerations or merited new Post Office facilities.

Liaison

Liaison is organised within the Post Office by sales and marketing staff who involve the appropriate engineering support groups as necessary. On the user side it is recommended that a single person or a single unit is responsible for ordering facilities from the Post Office and that the user nominates a local contact to deal with installation staff and to inform the ordering unit when the engineering work is complete.

Handover

The local user contact will be instructed by the installation staff how to operate the Post Office equipment, how to interconnect equipment, how to report faults and where necessary, how to operate remote tests on circuits or modems in conjunction with maintenance engineers. Operating instructions are being produced nationally for all Datel services and until they are available, details are often included in a locally produced aide memoire which usually also indicates relevant circuit designations, and the fault reporting point. Ideally a customer should be able to accept a circuit either by try out or test, while the installation staff are on site. Responsibility for the service remains with the installers for up to two days after commissioning to cater for any teething problems; thereafter, the appropriate maintenance staff take it over. Formal advice to the customer confirming the hand-over and the effective charging date complete the process some days later. Difficulties can and do occur but the important of good liaison between the customer's ordering unit and the local contact, as well as between the Post Office and the customer cannot be over-emphasised.

Advice

At telecommunications headquarters the Post Office has a data communications consultancy group. The function of this group is to provide advice on all aspects of data communications to users.

Connecting Non-PO Hardware

The user should only rarely become involved in obtaining permission to connect privately-owned and -maintained equipment. The responsibility to do so lies with the supplier. Copies of the Technical Guides and information on approved equipment are available from telecommunications headquarters at:

Post Office Telecommunications Headquarters
Service Department THQ
Tenter House
45 Moorfield
London EC2Y 9TH

THE NATIONAL COMPUTING CENTRE (NCC)

NCC Information Services

The purpose of NCC Information Services is to collect, verify, and maintain up-to-date information on all aspects of computing, and to be able to retrieve quickly that part of this information which is relevant to a particular enquiry.

The Information Services Division maintains files of information, classified at present into six Indexes.

Computing Hardware

Various details about computing equipment available in the UK are maintained in this index. Apart from computer systems, it includes peripherals, conversion equipment, storage devices, and data transmission equipment.

Computing Software

Information concerning computer programs known to be available within the UK is stored in this index. Generally these programs are provided by suppliers who are independent of computer manufacturers.

Computing Services

Details of software houses, consultants, computer bureaux, and other computing service industry organisations are included in this index, together with the individual facilities which they offer, and applications and disciplines in which they have particular experience.

Computing Education

This index contains information about the short term educational and training courses and facilities offered by computer manufacturers, consultants and other organisations. Details of conferences and seminars on topics related to computing are also included.

Computing Literature

This index contains abstracts of articles and technical papers published in computing periodicals and journals. Most of the abstracts are written at NCC, but other abstracting services, particularly in related disciplines, are taken to ensure complete coverage of the literature.

Computer Installations (The National Computer Index)

Within this index is held a record of computer installations in the UK giving the name and industry sector of the company operating the computer, the manufacturer and model of the computer used, the peripherals, the particular applications and other related information.

The information collected and verified at NCC is available either directly as prompt answers to particular enquiries, or more indirectly in the form of publications. For the former, enquiries may be made to Information Services by telephone, telex, letter or even personal visit.

As a consequence of experience gained operating the service some information products are produced regularly.

One is Computing Journals Abstracts. This is a fortnightly bulletin with an average of one hundred abstracts compiled from the current issues of more than 200 computing journals. This abstracting service available on an annual subscription provides subscribers with information concerning recent articles and papers and enables them to keep up-to-date with the advances in technology.

A monthly bulletin Computer Hardware Record contains all the standard abstracts relevant to a particular category of computing equipment. There is a different category considered each month. The outlay of the format of the abstracts in the bulletin make it easy to assimilate the information and to compare and contrast the different models. The subscriber to Computer Hardware Record will find it easy and convenient to prepare a short list of the equipment suitable for his particular and individual purpose.

Many clients request information from the National Computer Index in an ad hoc manner, specifying outputs to meet their precise requirements. These are produced from the National Computer Index using Filetab, NCC's own file handling and report generating package.

Within the Information Services Division regular routines operate to collect the information and maintain it up-to-date. These include contact with suppliers of equipment and services and with Members of NCC, and regular perusal of all issues of computing journals and press.

The NCC's Information Services are available to everyone whom they might assist. Where an answer in the form of standard abstracts is adequate, then organisations which are Members of NCC receive these free of charge. Enquiries from Non-Member organisations are equally welcome and are dealt with equally promptly, but are charged for the search of NCC's information indexes.

NCC Publications

The NCC publishes a range of books similar to this which aim to provide basic knowledge in data communications and computing. The most relevant of these are listed in the Bibliography. Full details may be obtained from any NCC office.

In addition, leaflets are available which cover such materials as the modems with Post Office approval.

MAINFRAME SUPPLIER

For information on software, communications control units, terminals and advice in general the mainframe suppliers should be approached via your organisation's usual contact in the computer company.

USEFUL PUBLICATIONS

The Computer User's Year Book

This annual publication gives useful information on the computer industry and on hardware and software suppliers in particular. It contains a fairly comprehensive list of data communications equipment suppliers, with a list of their general product ranges, contact address and telephone number.

Journals and Newspapers

Most equipment manufacturers advertise regularly in magazines and newspapers and this can often give a prospect buyer a feel for what is on the market. In addition articles and product reports from the same sources can provide an objective evaluation of various items of equipment.

Useful journals and newspapers include:

Computing
Computer Weekly
Electronics Weekly
Practical Computing
Which Computing?

OTHER USERS

In selecting data communications hardware a very useful source of information is other users. It is worthwhile discussing the matter with other members of your mainframe computer's user group if you are a member of one. Alternatively other users could be contacted via the data communications equipment suppliers you have invited to tender.

POST OFFICE TECHNICAL GUIDES

Technical Guide 1

Datel 100 Service. The connection of customers' equipment to Post Office Telegraph (including Telex) lines, circuits and equipment in the United Kingdom.

Technical Guide 2

General requirements for data terminal equipment.

Technical Guide 3

Datel 200 Service. The connection of customers' data terminal equipment to Post Office Datel Modems No. 2.

Technical Guide 5

Datel 600 Service. The connection of customers' data terminal equipment to Post Office Datel Modems.

Technical Guide 6

Datel 2400 Service. The connection of customers' data terminal equipment to Post Office Datel Modem No. 7.

Technical Guide 7

Datel 48K Service. The connection of customers data terminal equipment to Post Office Datel Modems No. 8.

Technical Guide 8

Automatic calling equipment. The connection of data terminal equipment to the Data Control Equipment No. 1A.

Technical Guide 9

Automatic answering equipment. Connection of data terminal equipment to the Post Office Data Control Equipment No. 2A.

Technical Guide 10

Data Control Equipment No. 3A. Automatic calling and answering equipment on Telex.

Technical Guide 12

PO Multiplexing Services, Dataplex System 2 the connection of Data Terminal Equipment to the Time Division Multiplexers 1A or 2A.

Technical Guide 14

Datel 200 Service. Connection of Data Terminal Equipment to the PO Datel Modem No. 13A.

Technical Guide 25

Transmission design requirements for the connection of privately owned and maintained equipment to Post Office private circuits.

Technical Guide 26

Protection design requirements for the connection of privately owned and maintained equipment to Post Office private circuits.

Technical Guide 30

Requirements for attachments connected to the public switched telephone network.

Technical Guide 32

General requirements and use of acoustically coupled modems for data transmission over Post Office lines.

Technical Guide 35

Requirements for attachments connected to Post Office and privately provided PBXs.

Appendix 4 CCITT V Series Recommendations

SECTION 1 – GENERAL

V.1 Equivalence between binary notation symbols and the significant conditions of a two-condition code

V.2 Power levels for data transmission over telephone lines

V.3 International Alphabet No. 5

V.4 General structure of signals of International Alphabet No. 5 code for data transmission over public telephone networks

V.5 Standardization of data-signalling rates for synchronous data transmission in the general switched telephone network

V.6 Standardization of data-signalling rates for synchronous data transmission on leased telephone-type circuits

SECTION 2 – INTERFACES AND VOICE-BAND MODEMS

V.10 Electrical characteristics for unbalanced double-current interchange circuits for general use with integrated circuit equipment in the field of data communications

V.11 Electrical characteristics for balanced double-current interchange circuits for general use with integrated circuit equipment in the field of data communications

V.15 Use of acoustic coupling for data transmission

V.16 Medical analogue data transmission modems

V.19 Modems for parallel data transmission using telephone signalling frequencies

V.20 Parallel data transmission modems standardized for universal use in the general switched telephone network

V.21	200-baud modem standardized for use in the general switched telephone network
V.23	600/1200-baud modem standardized for use in the general switched telephone network
V.24	List of definitions for interchange circuits between data-terminal equipment and data circuit-terminating equipment
V.25	Automatic calling and/or answering equipment on the general switched telephone network, including disabling of echo-suppressors on manually established calls
V.26	2400 bits per second modem standardized for use on 4-wire leased telephone-type circuits
V.26bis	2400/1200 bits per second modem standardized for use in the general switched telephone network
V.27	4800 bits per second modems with manual equalizer standardized for use on leased telephone-type circuits
V.27bis	4800 bit per second modem with automatic equalizer standardized for use on leased telephone-type circuits
V.27ter	4800/2400 bits per second modem standardized for use in the general switched telephone network
V.28	Electrical characteristics for unbalanced double-current interchange circuits
V.29	9600 bits per second modem standardized for use on leased telephone-circuits
V.31	Electrical characteristics for single-current interchange circuits controlled by contact closure

SECTION 3 – WIDEBAND MODEM

V.35	Data transmission at 48 kilobits per second using 60–108 KHz group band circuits
V.36	Modems for synchronous data transmission using 60–108 KHz group band circuits

SECTION 4 – ERROR CONTROL

V.40	Error indication with electromechanical equipment
V.41	Code-independent error control system

SECTION 5 – TRANSMISSION QUALITY AND MAINTENANCE

V.50	Standard limits for transmission quality of data transmission
V.51	Organization of the maintenance of international telephone-type circuits used for data transmission
V.52	Characteristics of distortion and error-rate measuring apparatus for data transmission
V.53	Limits for the maintenance of telephone-type circuits used for data transmission
V.54	Loop test devices for modems
V.55	Specification for an impulsive noise measuring instrument for telephone-type circuits
V.56	Comparative tests of modems for use over telephone-type circuits
V.57	Comprehensive data test set for high data signalling rates

V24 INTERFACE CONNECTIONS

Pin No.	Function	CCITT V24 Designation	Description of Function
1	Protective Earth	101	The Protective Earth is connected to the a.c. power earth and the modem frame.
2	Transmitted Data	103	Data from the Terminal Equipment that is to be transmitted.
3	Received Data	104	Received data from the modem to the Terminal Equipment
4	Request to Send (RTS)	105	The Terminal Equipment requests the modem to assume a Transmit condition by placing the 'Request to Send' line in an 'ON' condition
5	Clear to Send (CTS)	106	The modem provides an 'ON' condition on this line to indicate to the Terminal Equipment that it is ready to transmit data after receiving a 'Request to Send' signal.
6	Data Set Ready	107	The modem provides an 'ON' condition on this line to indicate that it is connected to line and all operate switches are in the correct mode to Transmit or Receive.
7	Signal Earth	102	Signal Earth is usually connected to Protective Earth inside the modem power supply.

8	Data Channel Received Line Signal Detector (Also referred to as a Data Carrier Detector)	109	An 'ON' condition indicates that the data carrier is being received.
9	Unassigned		
10	Unassigned		
11	Select Transmit Frequency	126	An ON condition indicates that the Backward Channel carrier is being received.
12	Backward Channel Received Line Signal Carrier	122	The modem provides an ON condition when the Backward Channel is ready to transmit data
13	Backward Channel Clear to Send	121	The modem provides an 'ON' condition to indicate that it is ready to send back Backward Channel Data.
14	Backward Channel Transmitted Data	118	Data to be transmitted on the Backward Channel
15	Transmitter Signal Element Timing	114	The modem provides a square wave signal at the clock frequency for synchronisation of the data to be transmitted from the data terminal
16	Backward Channel Received Data	119	Received Data on the Backward Channel

17	Received Signal Element Timing	115	This signal is a square wave at the bit frequency which is synchronised with the receiver
18	Unassigned		
19	Backward Channel Request to Send	120	The terminal equipment requests the modem to assume a Transmit condition on the Backward Channel
20	Data Terminal Ready	108	An 'ON' condition allows the modem to operate in the data mode when it receives some other stimulus
21	Unassigned		
22	Calling Indicator	125	An 'ON' condition indicates receipt of a call signal
23	Data Signalling Rate Selector	111	A Signal from the Terminal indicating the data rate to be transmitted.
24	Unassigned		
25	Unassigned		

DIFFERENCE BETWEEN PIN ALLOCATIONS IN RS232 AND EUROPEAN STANDARDS

Pin 11 is unassigned in RS232 but allocated in Europe to 'Select Transmit Frequency'.

Pin 21 is unassigned in Europe but allocated in RS232 to a 'Signal Quality Detector Signal'.

Pin 24 is also unassigned in Europe but allocated in RS232 to 'External Transmitter Clock'.

NOTE: This list gives the most commonly used interchange circuits. Between 8 and 18 of the V24 interchange circuits are necessary depending on the facilities required.

Appendix 5 Example of the Selection of a Portable Radio

a) Criteria by which radio is to be selected

- price
- size
- weight
- power supply
- connections
- controls
- tuning aids
- reception on VHF, medium wave and long wave

b) Rank criteria against each other

	Price	Size	Weight	Power Supply	Connections	Controls	Tuning aids	Reception
Price		2	2	3	3	1	2	2
Size	2		2	3	3	2	2	3
Weight	2	2		3	1	2	2	3
Power Supply	1	1	1		2	3	1	3
Connections	1	1	3	2		1	1	2
Controls	3	2	2	1	3		2	3
Tuning aids	2	2	2	3	3	2		3
Reception	2	1	1	1	2	1	1	
TOTALS	13	10	14	16	17	12	11	19

1 = less important
2 = equally important
3 = more important

Connections:	DIN output socket so that it is possible to record from radio + earphone socket + plugs into car aerial	10
	DIN output socket + earphone socket or plugs into car aerial	7
	DIN output socket only/earphone + plugs to car aerial	5
	Earphone only	3
	Plugs to car aerial only	3
Controls:	Sliding knob	10
	Just a few fixed positions	3
Tuning aids:	Automatic frequency control	10
	Station positions marked clearly on dial	4
	Nothing	1
Reception:	Judged on a sliding scale from 10 to 1, 10 being the best and 1 the worst.	

Features of models under consideration

	Model A	Model B	Model C
Price	£20 – £24.50	£30 – £34	£20.40 – £30
Size (cms)	14 x 21 x 6	18 x 30 x 7	19 x 32 x 9
Weight	0.9kgs	1.8kgs	1.9kgs
Power supply	Mains	Mains	Battery, mains adaptor
Connections	Earphone	Earphone	Car aerial, earphone
Controls	Few sliding positions	Variable know	Variable knob
Tuning aids	Stations marked	AFC	Stations marked
Reception	Fairly Good	OK	Good

Ranking/Score/Weighted Score for:

Criterion	Weighting	Model A			Model B			Model C		
Price	5	Cheap,	8	40	Fairly ex,	6	30	Cheap,	8	40
Size	3	Small,	10	30	Medium,	6	18	Medium,	2	6
Weight	5	Light,	10	50	Heavy,	5	25	Heavy,	5	25
Power supply	7	Mains,	1	7	Mains,	1	7	Choice,	10	70
Connections	7	Earphone,	3	21	Earphone,	3	21	Choice,	5	35
Controls	4	Bad	3	12	Good,	10	40	Good,	10	40
Tuning aids	3	Fair,	4	12	Good,	10	30	Fair,	4	12
Reception	9	Fairly good,	6	54	O K,	5	45	Good,	8	72
Total weighted score				226			216			300

Summary table for selection of a portable radio

Therefore Model C comes out as best by quite a considerable margin. This is primarily caused by it gaining high scores in two of the most important criteria, reception and power supply. It is interesting to note that in this case even if it had achieved a low score in one of these two Model C would still have come out as best. The buyer would then have to decide about a radio which comes out best on average despite the fact that it was the worst on reception or power supply.